With *Soju Party*, food writer, chef, and co-owner of Brooklyn's Orion Bar Irene Yoo has written the book on drinking like a Korean. She introduces the classic Korean alcohols and how Koreans typically like to drink them, including the viral Milkis Shot and a heart-stopping Seoul Train, and serves up unique cocktail recipes featuring Korean-inspired riffs and nostalgic twists, like a Jujube Ginseng Negroni and a Banana Milk makgeolli.

Of course, you can't drink without eating, and there are plenty of recipes for tasty anju (drinking foods), from simple snacks like the salty and sweet Honey-Butter Bar Nuts to essential comfort food like the savory White Ddukbokki and the super slurpable Kimchi Carbonara, with a dedicated party section featuring a large-format Watermelon Soju Hwachae and sweet-and-spicy Chimaek Chicken. In addition to recipes, Yoo explores the history of Korean drinking, with illustrations explaining proper serving and drinking etiquette, drinking games, food pairings, and more.

A book that promises late nights (don't worry, there's a section on hangovers!), this is a party on the page.

Geonbae!

SOJU

PARTY

SOJU PARTY

HOW TO DRINK (AND EAT!) LIKE A KOREAN

IRENE YOO

Photography by Heami Lee
Illustrations by Carolyn Yoo

ALFRED A. KNOPF   NEW YORK   2025

A BORZOI BOOK
FIRST HARDCOVER EDITION PUBLISHED BY ALFRED A. KNOPF 2025

Published by Alfred A. Knopf, a division of Penguin Random House LLC, 1745 Broadway, New York, NY 10019.

Knopf, Borzoi Books, and the colophon are registered trademarks of Penguin Random House LLC.

Library of Congress Cataloging-in-Publication Data
Names: Yoo, Irene, author. | Lee, Heami, photographer. | Yoo, Carolyn, illustrator.
Title: Soju party : how to drink (and eat!) like a Korean / Irene Yoo ; photography by Heami Lee ;
illustrations by Carolyn Yoo.
Description: First edition. | New York : Alfred A. Knopf, 2025. | Includes index.
Identifiers: LCCN 2024034452 (print) | LCCN 2024034453 (ebook) | ISBN 9780593802946 (hardcover) |
ISBN 9780593802953 (ebook)
Subjects: LCSH: Soju—Korea (South) | Cooking, Korean. | LCGFT: Cookbooks.
Classification: LCC TP607.S65 Y66 2025 (print) | LCC TP607.S65 (ebook) |
DDC 641.595195—dc23/eng/20241028
LC record available at https://lccn.loc.gov/2024034452
LC ebook record available at https://lccn.loc.gov/2024034453

Some of the recipes in this book may include raw eggs, meat, or fish. When these foods are consumed raw, there is always the risk that bacteria, which is killed by proper cooking, may be present. For this reason, when serving these foods raw, always buy certified salmonella-free eggs and the freshest meat and fish available from a reliable grocer, storing them in the refrigerator until they are served. Because of the health risks associated with the consumption of bacteria that can be present in raw eggs, meat, and fish, these foods should not be consumed by infants, small children, pregnant women, the elderly, or any persons who may be immunocompromised. The author and publisher expressly disclaim responsibility for any adverse effects that may result from the use or application of the recipes and information contained in this book.

The romanization of Korean words and phrases in this book follows the Revised Romanization of Korean guidelines and the online *Naver Korean-English Dictionary*. But given the lax implementation of Korean transliterations around the world, adjustments have been made based on readability, searchability of ingredients or locations, and the author's preference.

penguinrandomhouse.com | aaknopf.com

Food Styling by Eugene Jho
Prop Styling by Maggie DiMarco

Printed in China
2   4   6   8   9   7   5   3   1

The authorized representative in the EU for product safety and compliance is Penguin Random House Ireland, Morrison Chambers, 32 Nassau Street, Dublin D02 YH68, Ireland, https://eu-contact.penguin.ie.

FOR

엄마 and 아빠
(who don't drink)

NICK
(who definitely does)

and 제하
(who can't just yet)

우리가 소주를 마시는 이유가 단지 "맛 이 좋아서"만은 아닌 것 같다. 익숙해서, 어디서나 쉽게 찾을 수 있어서, 값이 싸서, 빨리 취하고 싶어서, 부장님이 좋아해서, 속 쓰린 일이 생겨서, 어려운 이야기를 꺼내야 해서, 분위기에 휩쓸려서, 맥주를 마시기에는 배가 부르고 위스키를 마시기에는 너무 비싸서, 회나 전골을 먹을 때 빼놓을 수 없는 술이어서 . . . 늘어놓자면 수십 가지 이유가 이어질 것이다.

—남원상, 『우리가 사랑하는 쓰고도 단술, 소주』

I don't think we drink soju simply because "it tastes good." We drink it because we're used to it, because it's easy to find, because it's cheap, because we want to get drunk quickly, because the boss likes it, because we're upset about something, because we have to talk about something difficult, because we got carried away by the vibes, because we're too full to drink beer and whiskey is too expensive, because we're eating raw fish or jeongol and can't not have it . . . I could list countless different reasons.

—Nam Won-sang, *The Bitter but Sweet Alcohol We Love, Soju*

Contents

2-CHA 2차
ANJU
NO PARTY WITHOUT AN ANJU PARTY 54

3-CHA 3차
KOREAN (MEETS) AMERICAN COCKTAILS
RIFFIN' ON THE CLASSICS

CHASING TASTE MEMORIES

4-CHA 4차
SOOL PARTY 술 파티

DRINKING GAMES 술 게임

5-CHA 5차
HANGOVER 해장
PARTY'S OVER (IT AIN'T OVER)

SOJU PARTY

INTRODUCTION

The text from my cousin Heedae pings through KakaoTalk: Injung-ah, do you want to go get a drink?

It's 11 p.m. I'm in Seoul, I've just returned from dinner, and I have a 9 a.m. flight back to the States to catch the next day. I should be packing, but I can't say no.

"Of course! Where at?"

Within the hour, I am sidling up to a mirror-backed bar near the Euljiro district, in a little second-floor speakeasy gilded with mother-of-pearl inlays, a cousin on each side of me.

After several Western-style cocktails—a Negroni for me, a gin and tonic for Heedae, a whiskey on one rock for his sister, Heejung—we decamp. It's 2 a.m. We've made short work of the basic catch-up ("How's your job?," "What happened to your last girlfriend?") and we need food and (cheaper) drink to dive into more. We pop our heads into a few different establishments, finally settling on a nearly empty pojangmacha.[*]

Our table is heavy with banchan,[†] raw fish, and miyeok guk[‡] by the time we polish off our first bottle of soju. As I order our second, I ask my cousins if they can teach me how to do the soju tornado.

Heedae, who's the same age as me, scoffs while Heejung, a few years younger than us, laughs. Our mothers are identical twins; I grew up spending every summer as a kid with them here in Seoul. They have both been drinking since they were nineteen years old[§] or probably younger. With condescending affection, Heedae crosses his arms to ask, "Why the hell would you want to do that?"

I pick up the sealed bottle that's just arrived at our table and give it a tentative shake; weak bubbles foam. Heejung grabs it from

---

[*] 포장마차—literally "covered wagon"; a street stall or pub, usually tented, that sells street food and alcohol. Also known by its shortened form, pocha 포차.

[†] 반찬—small side dishes that accompany rice, filling out a standard Korean meal

[‡] 미역국—seaweed soup, commonly served on one's birthday

[§] The drinking age in Korea is nineteen, according to a person's international age. Until 2023, Koreans followed their own age system, in which a person was considered one year old at birth and gained a year in age every January 1. The country now uses the international system, which calculates age based on the person's actual birth date (reducing everyone's age by one year) but follows their old system when calculating the drinking age. That is, a person can drink as of January 1 in the year they would turn nineteen. Confused yet?

me, overconfident, and attempts her own tornado. She, too, fails to achieve the desired swirl. Heedae, already red in the face (I'm pretty sure he began drinking before we met up), finally takes the bottle and shakes it aggressively, producing a violent funnel-shaped vortex.

"There's nothing to it," he says, shrugging. I am floored.

"Do it again!" I scream as I pull out my phone to take a video.

In the three hours that follow, we spill family secrets and wax on about never-ending work and the futility of life itself. Finally, at way past 5 a.m., I pour both cousins into a taxi, grab my bags from my Airbnb, and make the long schlep to the airport, my hangover starting to take shape. In line at security, I review my video of Heedae's tornado on repeat, bubbles twisting from cap to bottom like the warm fuzzy memories of our epic night in my brain.

I don't recall my first ever shot of soju—

where was I, who was I with?—but I remember so many nights of soju thereafter. Sometimes the drink is bitter, a bit painful. Sometimes it goes down like sweet water, like an afterthought, immediately requiring a follow-up shot to make sure everything is okay. I once patronized a rare Korean bar (among the countless Japanese izakaya) on St. Marks Place in New York City where I introduced my friend Uri to "the green bottle." We still talk about that night: the first shot, bracing, with each shot that followed tasting more and more powerful until, suddenly, it became nothing, flavorless. That is the danger zone: After we ordered one last bottle that we will never remember finishing, the film cut immediately to the next morning.

It takes me nearly a full year to gain the confidence, swagger, chutzpah, and exact flick of the wrist to achieve the perfect tornado. I practice every time I come across a soju bottle—at bars in New York City's Koreatown, across pochas in Korea, and during parties hosted in my apartment—to get it just right. My soju-lubricated friends cheer me on when I succeed and laugh consolingly when I don't. Then we get back to the business at hand, pouring shots for one another, dipping our spoons into spicy budae jjigae[*] and our chopsticks into sizzling platters of gopchang,[†] sharing stories, space, and heart-to-hearts.

[*] 부대찌개—army base stew, a spicy stew commonly made with Spam and instant noodles (see page 169 for my take)

[†] 곱창—grilled intestines

Much of my life has straddled two countries like this: my school years growing up in different parts of America before settling in a suburb of Los Angeles, and my summers spent with family in Seoul. Learning how to drink in South Korea was a slow education throughout my childhood—no blowout twenty-first birthday bash. Instead, I watched my uncles share a bottle of soju during family lunches, their faces and necks getting redder as they ordered another round and a plate of hwe,[*] from which my younger uncle would plop an iridescent slice of amberjack swiped in chojang[†] into my mouth. "This is how you eat when you drink." I helped pour a bowl of makgeolli at my family's charye[‡] table as my umma whispered the correct order into my ear: "Your oldest uncle first." I sipped a beer for the first time in my appa's lap. "Here, have a shrimp cracker to cut through the bitterness."

Alcohol—and the food eaten with it—has always been a way for me to connect and share with others. When a party I'm hosting starts winding down, I ask, "Who's hungry?" I put on water to boil, open a dozen bags of instant ramyun, and set the steaming pot on the table for everyone to feed themselves. It's my favorite part of the night, watching my friends, drunk and hungry, happily eat something I've made them before we settle in for more.

In 2015, I started popping up at bars around Brooklyn as "Yooeating," serving Korean-themed menus designed for snacking on alongside a cocktail or two. With my now-husband Nick, I hosted dozens of these pop-ups: cooking bowls of noodles, pressing gooey cheesy sandwiches, even making pizza and lighting spicy ddukbokki[§] on fire. In 2024, we opened Orion Bar in Bushwick with the goal of creating a new style of Korean and American drinking culture, something in between a K-town pocha and an American dive bar. Since then, we've been serving somaek,[¶] our own Korean twists on classic American cocktails, and my crave-worthy bar snacks and late-night ramyun. I want

[*] 회—raw fish

[†] 초장—a spicy-sweet dipping sauce made with gochujang and vinegar

[‡] 차례—a memorial service for ancestors observed on anniversaries and major holidays like Chuseok

[§] 떡볶이—Korean rice cakes

[¶] 소맥—soju energy bombs

to share the Korean communal style of drinking and the comfort food I love in the same way that it has been shared with me.

Drinking is a way of life in Korean society: You drink with your family to show respect for your elders and ancestors, you drink socially with your friends, you schmooze and booze with co-workers and clients. All this drinking comes with an implicit set of rules in our culture: how to first receive a drink, then how to pour one. Most Koreans learn these rules by sharing a table with someone older than them, but maybe you've never had a Korean drinking mentor (and, even if you have, the more drinking buddies, the merrier!).

The alcohol that Koreans drink—soju, beer, and makgeolli—contains less than half the alcohol of stronger liquors like vodka. This is precisely what shapes the uniquely Korean way of drinking. By knocking back shots and drinking in smaller doses, we're able to drink for longer, until the wee hours of the morning. Another key element of our drinking culture is the tradition of pairing alcohol with food. Anju 안주 is a category of dishes that's specifically served as an accompaniment to each drink as well as a way to absorb the alcohol. For me, the perfect "chaser" isn't beer or soda—it's a spoonful of spicy clam soup after a particularly sharp shot of soju.

So with this book, I want to be your guide. Your elder, who will pour you that first drink. Your hard-drinking noona[*], who will go shot for Milkis shot (see page 20) with you, demanding that you keep up. Your friendly bartender, serving up unique cocktails reflective of my Korean American childhood. Your unni[†], who will take care of you and make you a big pot of instant ramyun back at my place after a night of drinking. Your friend, who will teach you all the fun party games the night before and then have hangover soup with you the morning after. I'll share with you how to drink, how to pace yourself, how to pair the best foods with all those drinks, and all the other things I've learned. We'll get convivial and generous, as Koreans do when we get drunk. Our formalities and unspoken judgments will drop. We'll clink glasses and suggest another round, morphing from strangers into lifelong friends. Geonbae!

[*] 누나—an older sister, to a male

[†] 언니—an older sister, to a female

How to Use/ Drink Through This Book

Whenever I'm drinking "Korean-style," I know I'm strapping in for multiple rounds. We call this "1-cha, 2-cha, 3-cha" (1차, 2차, 3차), or simply "first, second, third." Each round provides a change of scenery and an opportunity to switch up the direction of the night, as well as offering natural exit points for anyone who might want to drop out.

HERE'S WHAT A
TYPICAL NIGHT
MIGHT LOOK LIKE

1-CHA 1차 We meet up at a **makgeolli bar** for some **light sips** and **bites**. It's an opportunity to get the small talk (and our buzz) going while we wait for any laggers who might be wrapping things up at work.

2-CHA 2차 We head to **dinner**—maybe a hot-ticket samgyeopsal (pork belly BBQ) restaurant where we don't mind waiting for a table since we've already got a few under our belts. We order our **first bottles of soju** and make short work of them while we grill our fatty meats.

3-CHA 3차 We decamp to a **cool cocktail bar** or a **hof** (see page lviii) where we can ramp up the alcohol ante. We're served some **small bar snacks**, and I can never resist ordering a plate of half-dried squid to gnaw on.

4-CHA 4차 If our energy level still feels high, we'll park our butts in a **coin karaoke room** with a few more bottles of soju and sing our hearts out. If we're feeling a little too boozed, we'll do **dinner, round two**—either chadolbaegi (beef brisket BBQ) or straight to a bowl of three-in-the-morning haejang guk (hangover soup).

5-CHA 5차 Last round at home! My favorite nights out in South Korea, after going round after round at different clubs, pochas (see page lix), or bars, end with **one last drink at home**. We'll stop by the 24-hour convenience store to grab snacks and something fun to drink, like a limited-edition mint chocolate soju (surprisingly good!). We'll kick off our shoes, recap the night, dance to our favorite tunes, and fall asleep on the couch until 6 a.m., when we'll be sober enough to wash up and make it back to our actual beds. This is when that last pot of drunken ramyun is key.

THE CHAPTERS IN THIS BOOK
ARE STRUCTURED LIKE
A NIGHT OF DRINKING

For the first round (1-CHA), we'll ease into Korean-style drinking with an introduction to the classic Korean alcohols and how Koreans typically like to drink them.

I'll make sure you get fed during 2-CHA with tasty anju that's made for eating when drinking, with snacks like salty and sweet Honey-Butter Bar Nuts (page 62), dishes like the savory and just-a-wee-bit-secretly-spicy White Ddukbokki (page 90), and, of course, my super slurpable Kimchi Carbonara (page 97).

At 3-CHA, we'll level up to more complicated drinks featuring Korean-inspired riffs on classic cocktails, many of which can be found at Orion Bar.

We'll peak at 4-CHA, entering full party mode with a large-format Watermelon Soju Hwachae (page 151), an iconic Samgyeopsal spread (page 157), and some mouthwatering Chimaek Chicken (page 161) that's perfect with beer.

And, lastly, I'll help you ease into your hangover during 5-CHA, with recipes that you can whip up after a long night out.

Korean Drinking Traditions

Whether drinking while out with friends, at dinner with family, or during a celebration for the boss's promotion, Koreans follow rules that dictate how to do it properly. Some of these formalities stem from our culture's history, some are holdovers from now-outdated practices, some my uncle probably made up.

Koreans greatly emphasize respect for elders within family and group structures, largely due to the Confucian influence that underlies the culture. A parent's word is law, especially since many children continue to live with their parents well into adulthood. You're expected to defer to the wisdom of your parents and the seniority of your bosses and generally display a proper level of respect for anyone older than you. Respect is also shown physically, by bowing in greeting and affirmation, and is integral to the Korean style of drinking, where there are clearly established (though unspoken) rules around who orders, when to order, how to pour, who pours, and on and on.

These traditions stem from ancestor worship, where alcohol ritual is especially foundational. Historically, Koreans host rites to honor the deceased on anniversaries or holidays like Chuseok. This is something my family and I practice, too, both in South Korea and America: sprinkling the contents of a bottle of soju on mountainside grave mounds in Gimpo, passing small cups over incense on the ancestral worship table for the anniversary of a great-aunt's death in Glendale, and drinking and eating the prepared foods with one another afterward. During these rituals, a specific pouring order is observed. The eldest son will pour the first cup, then the second eldest, and so on, a hierarchy of pouring that carries over into everyday drinking settings as well.

Today, drinking plays an outsize role in Korean work culture, where office workers are expected to accept every invitation to drink, and to drink until they have to be scraped off the floor and carried into a taxi. Pouring and drinking hierarchy is crucial here: You pour for your boss, they pour for their boss above them.

But it's not just at work; much of Korean drinking centers around establishing hierarchy. In more friendly settings, this is loosely based on age—you pour for those older than you or in school grades above you (this lasts well after graduation). In first-time social settings, the hierarchy becomes even stricter and more formal. If an age hierarchy is not immediately apparent, there's often an introductory sharing of birth years to establish a loose rank.

For me, part of the fun of drinking soju is rehashing these rules and rituals as a nod to those who taught me to drink and passing them onto others. Choose the ones you like, forget the ones you don't. It's like a good tale told many times over: The words may change, but the heart of the story carries on.

SOOLJARI 술자리
(KINDS OF DRINKING SITUATIONS)

With Your Family

Ideally, your most forgiving opportunity to practice your soju drinking skills. Accept every drink your older family members offer you, and partake heartily in all the anju they order. Be respectful and don't get too sloppy, but bank your uncle's drunken imitation of the latest K-pop dance craze in your memory.

With Your Friends

In Korea, friends, or chingu 친구, are generally people who are the same age as you. That means there's a connection to start, and you can drink together in a more casual and relaxed way.

With a Mixed Group

Out of respect, treat everyone as if they are older than you to start, until you've had a few drinks and established clear relationships.

With Your Boss or Clients

Stay on your toes and always maintain the utmost respect; this is where Korean hierarchy is most important. Observe pouring rules religiously (see How to Pour, page xxvi).

With Your Co-Workers

Once your boss or the important client leaves (yay!), decamp to a different location so you can properly vent about work and let loose.

By Yourself

When you're drinking alone (also known as honsool 혼술), all bets are off, but try and maintain a little dignity. Pour your soju into a glass instead of drinking straight from the bottle and grab a small bag of snacks so you're not drinking on an empty stomach. (Who's gonna take care of you but you?!)

Note A lot of these rules, particularly those pertaining to pouring and receiving, are applicable to all Korean drinking, but they tend to apply most to soju drinking, which is inherently more formal.

HOW TO **POUR**

Never pour for yourself. Instead, pour in order: Since Korean society is sensitive to hierarchy and signs of respect, the ritual of pouring is extremely important. Usually, soju is poured for the most senior person at the table first, then for the next most senior, and onward down the line until the lowest assistant or youngest person is served. There may be a natural switch if your place in the hierarchy comes up, or someone else volunteers to continue pouring in your stead.

There are instances where whoever is buying will initiate the pouring, particularly if there is someone who will clearly be paying for the meal, since it's technically their alcohol to serve. Give it a beat when the alcohol arrives at the table and see if someone reaches to pour first, which they will especially if they are big drinkers. Similarly, if you don't know where you rank in the group, offer to pour for whoever appears to be the eldest or the most senior, or wait until you are offered a glass and then offer to return the favor to your pourer.

These rules can be a little more relaxed in a friend setting. For the first round, it's customary to follow the rules according to age (if known) or according to perceived hierarchy (i.e., pour for the most esteemed or most unfamiliar guest first). In consecutive rounds, as present company gets more relaxed or you learn one another's ages, this hierarchy can shuffle or be dropped.

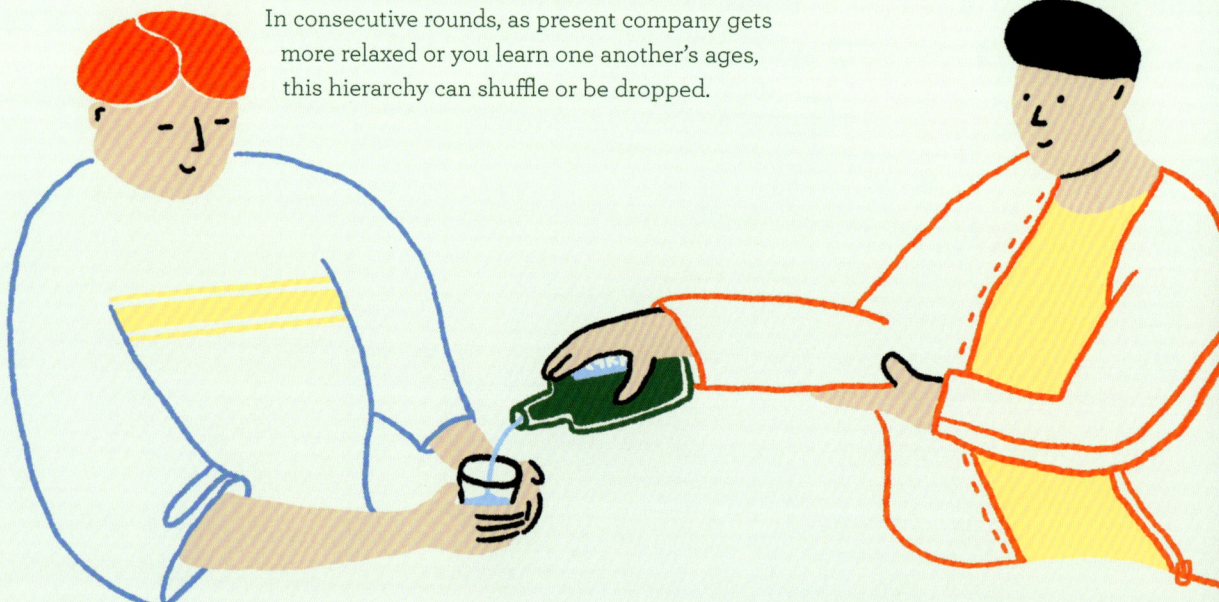

How to Hold the Bottle Wrap your right palm (even if you're left-handed) around the front of the paper label. Covering the label dates back to when label printing was poor and the ink would run when wet. Obscuring the brand or region also served to avoid offending the receiver if they preferred a different brand or hailed from a different region. In any case, it's more of a tactic to grasp the bottle properly—don't balance it from the bottom like a sommelier serving wine or choke up too far on the neck. Place your left hand gently at the base of the bottle or under your elbow as you pour, or under your armpit if you're reaching quite far over the table.

Don't Pour Your Own Glass Once you've finished pouring for others, place the bottle down, or subtly pass it to the person you last poured for. Do not pour your own glass! Wait for someone else to pour for you.

How to Receive a Glass Hold your glass off the table with both hands, or with your right hand while your left hand gently supports the bottom of the glass or your elbow.

The Polite Finger This is something I've seen more recently, particularly from my cousin Heejung, who is unfailingly polite in all settings. When someone is pouring for another person, she will reach over and place a polite finger on the edge or side of the recipient's glass, indicating that she is respectfully offering this pour as well. You can execute the same movement if someone is pouring their own glass (which gets more likely as the night wears on and you're not keeping tabs as much). You can do this with an outstretched palm as well, lightly hovering or touching the tip of your hand to the receiver's glass, as a greater sign of respect.

HOW TO **DRINK**

How to Cheers Tap your glass gently with your tablemates'
when toasting—even here you must respect the hierarchy! Keep
your glass slightly lower than the glasses of those older or more
senior than you, maintaining the same grip you had when
receiving the glass. You can also accompany the clink with a
slight bow to communicate additional respect.

Turn to the Side When drinking with people older or more
senior than you, or in unfamiliar company, turn your head and
eyes away from them to take your shot. Use your hand to politely
cover the side of your mouth and glass that faces the table as you
do so. Like bowing, split the difference between too little and too
much; there's no need to turn your back fully to the table.

As the drinking continues, keep your eye on other people's
glasses, and offer to refill if you see them empty. If your own
glass is empty, offer to pour someone else's glass, which should
prompt them to then fill yours. Don't worry, someone's got you
covered.

An Introduction to Korean Alcohol

Korean drinking revolves mostly around three types of alcohol: soju, maekju (beer), and makgeolli. While many Koreans do drink a wide variety of alcohol—especially now, when foreign liquor has become quite popular—these three are the most widely produced and consumed. Soju reigns supreme, having evolved from a traditional distilled liquor to the commercial green bottles that litter every pocha,* convenience store, and street corner in South Korea. Beer plays a major supporting role, as a go-to mixer for soju and a refreshing low-ABV drink for after-work imbibers. And makgeolli, the original Korean alcohol, has enjoyed a major revival among microbrewers and discerning drinkers. I'll dive into how each of these came to represent Korean drinking today, and touch upon the other regional and niche liquors within the country's rich drinking landscape.

* Short for pojangmacha (see page lix)

Soju

소주

Whenever I drink soju, I squirrel away the empty green bottles in the corner or under the table at my feet (if the servers don't steal them away from me first). Maybe it's a helpful way to track how much we've drunk as the clinks of shot glasses start to blur together, but in truth, I just like the sound of the glass knocking together, the bottles piling up as the night wears on. Each bottle contains the remnants of our joys or our sorrows, a friendship forged, a story told.

Soju has a story to tell, too. It's by far Korea's top consumed alcohol, drunk at every bar, restaurant, and convenience store regardless of the time of day. The most popular and commercially available version of this soju is nearly always sold in a 360 ml green bottle (375 ml here in America), with an alcohol content somewhere between 12% and 20% (a number that's been decreasing over the past few decades), and generally tastes the same no matter the company or the brand. They all share the same nose-tickling ethanol burn, followed by a mellow sweetness that peters out on your tongue, leaving you thirsty for more.

But everything about this soju has evolved significantly from its original iteration, which dates back to the country's early formative years: the flavor, the alcohol, the production method—even its green bottle.

The Start of Soju

The story of soju begins early in Korea's history. The Mongols, who ruled China through the Yuan dynasty during Korea's Goryeo era (1270–1356 AD), relied on alcohol as a tool to boost troop morale and spread political influence. One of the drinks they popularized was arak, an Arabic distilled liquor from the Middle East made from fruit wines or fermented milk. High-proof with a long shelf life, it was easy to travel with into China and Korea.

Once there, arak evolved into early versions of Chinese shaojiu (known today as baijiu) and Korean soju, both made with widely available local grains like sorghum and rice, which were fermented and then distilled. The simultaneous evolution of these alcohols is evident in the etymology of their names; the word *soju* shares the same Chinese characters as *shaojiu,* which means "burnt liquor."

The Mongols further influenced Korean alcohol production by improving on an existing form of Chinese distillation to make a portable still, the soju-gori 소주고리 (or sojutgori 소줏고리), a vessel usually constructed out of clay, but sometimes copper or brass. It featured two compartments stacked on top of each other, separated by a narrow opening, with a spout coming off the top compartment. To make soju using a soju-gori, rice was first fermented, using nuruk 누룩 (a fermentation starter) and water, into an unstrained rice wine called takju 탁주 or a clear strained wine called cheongju 청주 (also yakju 약주). The takju, cheongju, or both combined (called wonju 원주) were placed in the lower chamber of the soju-gori and boiled. The evaporated alcohol would rise to the upper chamber, where it was cooled with cold water and returned to liquid form, which flowed out the spout as distilled soju. These soju-gori were compact and easily movable, so many people during the Goryeo era learned to distill their own alcohol at home, creating a home-brew culture known as gayangju 가양주.

Now that people could easily make their own, soju spread rapidly during the following Joseon dynasty (1392–1897) and became ingrained in Koreans' daily lives. It was served to visiting guests, poured during ancestral rites, and infused with flowers like chrysanthemum or herbals like jujubes or ginseng to make medicinal liquor. The government actively encouraged this home brewing, taking soju as a form of tax, instead of building its own breweries or distilleries. Soju also became popular in the royal courts, where nobles enjoyed sharing it as ritual offerings abroad and drinking copious amounts at home.

ISN'T SOJU JUST KOREAN SAKE?

Absolutely not! For starters, sake is a rice wine that's brewed, not distilled. The closer comparison would be Japanese shōchū, which is a clear distilled liquor. Soju and shōchū share a similar etymology ("burnt liquor"), but the history of soju is older—Koreans developed the distillation technique and helped bring it to Japan.

The Diluted Soju Industry Emerges

The Japanese occupation of Korea (1910–1945) had lasting effects on many facets of Korean culture and society. Koreans were banned from using their own names and language, forced to adopt Japanese names and speak only in Japanese. Korean agricultural resources, like rice, were co-opted and exported to Japan, and many Japanese companies set up businesses and factories on Korean soil, both to expand production and to feed the local Japanese population.

The introduction of Japanese-style food production profoundly affected how traditional Korean food was made. For example, Korean-style soy sauce (ganjang), traditionally brewed at home by individual households, was largely replaced when Japanese soy sauce companies built factories in Korea. The process was industrialized, making it cheaper to manufacture and buy. Despite clear differences in taste and quality, this new style of soy sauce became so mainstream that most Koreans today no longer recognize the flavor or style of the traditional Korean version.

In the same way, Japanese influence changed traditional soju into a modern, more mass-produced version of itself. First, the Japanese introduced the continuous still (also known as a column still), which had been brought to Japan by Europeans in 1895. It served as a key component of industrialized alcohol production, increasing the ability to distill alcohol at much larger volumes and at much lower costs.

The ingredients used to make soju—originally just nuruk, rice, and water—changed, too. Nuruk was replaced by Japanese kōji, a more stable and controllable fermenting agent used for making soy sauce and miso. "Black kōji," a Japanese yeast made with black beans, was later introduced as an even cheaper substitute, and the use of nuruk, utilized in the production of 99 percent of soju in 1923, fell to 5 percent by 1932. Since rice was being exported for Japanese consumption, it, too, was slowly replaced, with cheaper sweet potatoes and tapioca. (They did keep the water, though!)

WHAT'S THE DIFFERENCE BETWEEN NURUK AND KOJI?

Nuruk 누룩
Korean wild-fermented milled wheat/grain cakes

Nuruk is a traditional Korean fermentation starter, used to kick-start makgeolli and traditional soju. It usually comes in dried round disks made from grains, such as wheat, rice, barley, or a combination, that have been fermented in open air or in a heated room called an ondol. These disks contain enzymes and yeast that, when mixed with rice and water, break down the starches into fermentable sugars, creating alcohol. Stateside, you can find nuruk at your local Korean grocery store, sometimes enzyme amylase, sold in a bagged, more crumbled form.

Kōji
Japanese *Aspergillus oryzae* cultivated on whole rice

Kōji is a strain of the fungus *Aspergillus oryzae*, which is intentionally applied to rice to kick-start the fermentation of alcohols like sake or shōchū, or condiments like miso and soy sauce. While kōji is a purposeful, controlled inoculation, nuruk is more of a wild fermentation that naturally produces the same *Aspergillus oryzae,* in addition to many other yeasts and bacteria, such as lactobacillus.

The production of soju was further industrialized following the Korean War in the 1950s. As the South Korean government struggled to feed its population and rebuild its economy, it introduced new laws that banned traditional home-distilling methods and the use of rice in alcohol production. It was at this point that the now-standard diluted soju first appeared: Instead of distilling soju directly, a high-proof ethanol made from sweet potato or tapioca was mixed with water, sweeteners, and preservatives, a far cry from traditional soju.

Competition and diversity in soju-making was also greatly reduced at this time. In the 1970s, the government forced each South Korean province to choose one major soju company to represent its region, effectively consolidating or shutting down smaller producers. The number of soju companies shrank from 340 prior to 1973 to just 13 by 1977. You can see this regional reduction still represented today: the brand Daesun represents Busan, Hallasan represents Jeju Island, and so on.

SOJU BY REGION

COMPANY	FOUNDED	REGION	PRODUCTS
HiteJinro 하이트진로	1924 as 진천양조상회	Seoul/ Gyeonggi-do	Chamisul 참이슬, Jinro Is Back 진로 이즈 백
Lotte Chilsung 롯데칠성음료	1926 as 강릉합동주조	Seoul/ Gangwon-do	Chum-Churum 처음처럼
Muhak 무학	1929	Gyeongsangnam-do	Good Day 좋은데이
Kumbokju 금복주	1957	Daegu	Charm 참
Bohae 보해양조	1950	Jeollanam-do	Yipseju 잎새주
The Mackiss Company (also known as Sunyang) 선양소주	1973	Daejeon	O_2Linn O_2린 (now Malgeulinn 맑을린)
Daesun Distilling 대선주조	1930	Busan	C1, Daesun 대선
Chungbuk Soju	1957	Chungcheongbuk-do	Siwon (Cool Cheongpung) 시원한 청풍
Hallasan Soju	1950	Jeju Island	Hallasan Original 한라산

The government also monopolized the production of the high-proof ethanol used to make soju—that ethanol is still made by just one manufacturer today—and began to control its distribution to these companies. This ultimately created the current situation where almost all soju made in Korea is produced in the same way, and the soju companies (what few are left) use marketing and price wars instead of taste to differentiate themselves from one another. It's why modern soju is so unique in its sameness.

Why Is Soju Sold in Green Bottles?

Before I was even old enough to drink, I used to wonder why all soju came in the same green bottles. Was it a government mandate to streamline recycling? (But then why only soju?) Did all the soju companies get together to agree to use the same bottles? Is this all just a conspiracy and all soju is actually the same?!

Turns out, before the 1980s, soju bottles came in many different colors: blue, brown, or clear. In 1982, the South Korean government began to standardize the bottle colors to differentiate between soju sold for home consumption (brown) versus those sold through businesses (blue) so that they could charge higher taxes on the latter. When this endeavor proved unsuccessful (businesses would just secretly buy the brown bottles to circumvent taxes), in 1984 all soju bottles became clear blue, which also helped to reduce manufacturing and recycling costs.

But the color matter wasn't settled. In 1993 Dusan Gyeongwol 두산경월 launched Green Soju 그린 소주, the first soju to be sold in a light green bottle. They wanted their soju to seem more "natural," clean, and easy to drink. It was a hit, blowing the previously number-one soju producer, Jinro 진로, out of the water. In 1996, Jinro eventually designed its own green bottle to regain its position on top.

In 2009, most of the soju companies voluntarily agreed to adopt the most popular shape, color, and size to streamline recycling. The bottle alone accounts for 30 percent of the cost of making soju (and it's a pretty cheap alcohol to begin with), so this change would have a huge effect on profit margins. They chose Jinro's green 360 ml bottle as the new standard, which is still used today. Now green bottles can be interchangeably returned to and reused by nearly all soju companies.

With one exception . . .

Although almost all soju now comes in green bottles, the iconic Jinro Is Back, introduced in 2019, features a *blue* throwback design that capitalizes on the "newtro" (new-retro) trend that's popular in Korea. Its rounder, squatter shape and translucent blue color are specifically designed to stand out from the standard green bottles.

As sales of Jinro Is Back increased, problems in the recycling pipeline arose. When blue bottles began flooding their recycling streams, HiteJinro's main rival, Lotte Chilsung, initially refused to return them for months. Eventually, a new agreement was reached where the companies would trade the blue bottles back in exchange for the same number of standard green bottles.

Jinro Is Back is technically not the first soju to break the mold; Hallasan (of Jeju Island) has long been sold in clear bottles—and Lotte recently launched their own non-green soju, Saero 새로, a zero-sugar soju in a clear, lightly ridged bottle.

At this rate, maybe green-bottle soju won't even be the standard in a few decades!

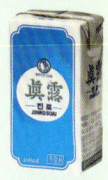

A SOJU OF ANY OTHER SIZE

As iconic as the green bottles are, I'm also particularly enamored with the other types of packaging on the market. The smallest packaging that soju comes in is a 160 to 200 ml paper juice box, known as pack soju (팩소주). This was introduced by Jinro in 1989 and marketed as a convenient way to imbibe while participating in leisure activities like fishing or watching baseball (they're made to slip right into your pocket!). There are also plastic 200 ml pocket flask–sized bottles for easy travel (my friend's mom brought twenty bottles to tide her over on her recent trip to New York from Korea, one for each night). On the larger side, you can find PET bottles of soju ranging from 400 ml to 5 liters, targeted for high-consumption usage for businesses, college retreats, and home alcohol-making.

Where Did the Alcohol Go?

Koreans love to ask each other what their tolerance is based on the number of (green) soju bottles they can drink in one night. For most, it's one bottle. For the mighty, it's two. And for the heavy drinkers, well, they stop counting after the second one.

But historically not all bottles of soju are equal, something my uncles will wax poetic about as they pour themselves another glass. "Yah, back in the day soju was so much stronger, not the swill that kids these days are drinking."

And it's true. Initially, traditionally distilled soju had an ABV of 40% or higher, but with the industrialization of soju during the Japanese occupation, the alcohol percentage started to decrease. Machine-distilled soju produced during the Japanese occupation clocked in at 30% to 40%. The alcohol dropped a bit in the decades following the occupation and the Korean War as companies moved to making diluted soju, but this soju was still as close to 30% as possible; anything less had the reputation of being "watered-down soju" 물탄소주.

In the 1970s, Jinro lowered the ABV of their top-selling soju to 25% due to reduced access to ingredients. Crucially, they dropped the price as well, and sales actually increased by 50 percent. Other companies quickly followed suit, and by the 1990s, the floodgates opened in the race toward lower and lower alcohol percentages.

Initially the lighter version was marketed as being fresh and smooth, like a drop of dew or mountain water. And through financial crises and economic recessions soju companies kept the price of their products appealingly accessible by continuing to quietly make them less and less alcoholic. Soju became more refreshing and drinkable, while the bottles became—and stayed—cheaper, and the lower ABV had the added effect of making everyone drink even more!

A RACE TO THE BOTTOM: THE REDUCTION OF ALCOHOL CONTENT IN KOREA'S TOP-SELLING SOJU

	CHUM-CHURUM	CHAMISUL
1998		23%
2006	20%	20.1%, 19.5% (as Chamisul Fresh)
2007	19.5%	19.8%, 19.6% (as Chamisul Fresh)
2012	19%	19.5% (Chamisul Fresh)
2014	18% (in February), 17.5% (in December)	18.5% (back to Chamisul), 17.8% (Chamisul Fresh)
2018	17%	17.2% (Chamisul Fresh)
2019	16.9%	17% (Chamisul Fresh)
2020		20.1% (Original), 16.9% (Chamisul Fresh)
2021	16.5%	16.5% (Chamisul Fresh)

Note Manufacturers often adjusted the ABV of their products multiple times within the same year, so this chart is not meant to be an exhaustive list of changes.

You can see how quickly the ABV plummeted—nearly 7% in just over fifteen years. That's more alcohol per serving than most beers!

ARTISANAL SOJU: A RETURN TO TRADITIONAL METHODS

Traditionally distilled soju began to experience a slow resurgence in Korea leading up to the 1988 Summer Olympics, when the government encouraged artisans in regions like Andong to make and preserve this old-style liquor. As the economy improved and restrictions around alcohol-making loosened, the big soju companies started getting into the game (HiteJinro's premium soju line Ilpoom Jinro launched in 2007) and small-batch distilleries began to crop up, like Sulseam, established in 2012.

In 2016, my friend Brandon "Bran" Hill started making his own soju, inspired by his study of traditional distilling in Korea. At the time he was the master distiller at Van Brunt Stillhouse, a Brooklyn-based whiskey distillery. He started playing around with making soju and makgeolli in his spare time, eventually launching Tokki Soju, the first US-made soju, with his partner Douglas Park. They sourced nuruk, the Korean wild fermentation starter that's hard to find and even harder to distill with consistently, and used only rice and water, iterating laboriously to get the flavor profile and methods just right. The soju Bran made was strong (at 40% ABV) but tasted so good; we served shots of it at our wedding.

Tokki's success in the States began generating buzz in Korea, so Bran and Doug moved their distilling operations to Chungju in 2019 and have since expanded globally. And they're not the only ones launching new types of soju. In 2015, Korean American Carolyn Kim created Yobo Soju, making soju distilled with grapes, as well as aperitifs and flavored soju, while Dan Lee and Max Fine started West 32 Soju (named after New York City's Koreatown) in 2016, producing their soju from distilled corn. Korean rapper Jay Park famously launched Won Soju in 2022, his answer to the American style of celebrity-led alcohol brands. As the interest in soju grows worldwide, soju continues to break out of its green-bottle mold, with new styles of artisanal and commercial soju being introduced.

Same Same but Different. But Still Same.

Soju is almost flavorless, so it can be difficult to differentiate between different brands and products.

 **Chamisul** Fresh and Original are lighter in flavor and a little less sweet than their sister soju Jinro Is Back. Fresh is very light and thin, almost watery, while Original has more of a biting alcohol burn.

 **Chum-Churum** Less watery than Chamisul Fresh. It starts with a bit of a harsh alcohol note that mellows out into sweetness.

 **Jinro Is Back** Has a bit more flavor, more viscosity, and a little more sweetness than Chamisul. It carries a nice evenness throughout.

 **Charm** Similar to Chamisul Fresh, it tastes quite watery.

 **Hallasan** Very clean with less of the sweet aftertaste.

 **Daesun** Nicely balanced and pairs well with food.

 **Tokki Soju** Its sweetness tastes more like rice than sweeteners, with more of the distilled alcohol flavor that you might get from a vodka or unaged whiskey.

 **Ilpoom Jinro** Even though it has a higher alcohol content, it tastes nearly as smooth as the lower ABV soju.

FLAVORED SOJU

I personally am not a fan of flavored soju. When I'm just looking for a regular green bottle, I don't want to be distracted by blueberry or grapefruit or even yogurt-flavored versions. But I get it: It's something younger, college-age kids in Korea often first reach for when learning how to drink (I went through that phase, too. It's like peach schnapps or banana-flavored rum: a sweet entry point that may seem off-puttingly cloying once you get into the real stuff.)

In the 1980s and 1990s, Gen Xers enjoyed drinking at "soju-bang" 소주방—think a karaoke room without the karaoke. These became popular for dates and groups of women, serving "cocktail soju," a soju-based concoction mixed with fruit juices like cherry, lemon, or orange (or a combination) and topped with soda. This spurred soju companies to productize cocktail soju for a brief time in 1994, later reintroduced in 2015 as "fruit soju."

Today's bottled fruit soju builds on an already heavily chemicalized soju formula; the fruit flavors often taste like cough syrups or cheap candy. But soju with real fruit juice or fresh fruit is still a great idea! Check out my Watermelon Soju Hwachae (page 151) or my Berry Ju (page 45), as proof!

So What's the Real Soju?

Given the roundabout trajectory of soju's history, I sometimes find myself staring down the barrel of a green-bottle soju, filled with a chemical approximation of the traditional distilled version, and wondering, "Are you even really soju?" What about these newer brands, which return to older recipes and methods but incorporate new ingredients or techniques?

Why not both? Diluted soju (aka green-bottle soju) and distilled soju coexist today, and Koreans have never stopped drinking soju as it's evolved through its many forms. Even though bans on using rice in alcohol-making and regulations controlling alcohol levels were lifted by the late 1990s, green-bottle-style soju continued to grow in popularity. For older generations, it may have been the only soju they could access or afford, and for newer generations it might be the only soju they've ever known. It's everywhere—convenience stores, K-dramas, pochas, as well as liquor stores and Korean BBQ restaurants here in America—becoming cheaper and more accessible out of both necessity and popularity. Meanwhile, distilled soju is now treated more like a premium alcohol, made for thoughtful sipping or incorporating into cocktails. Even though the original formula may have been lost or changed during times of struggle, soju evolves and perseveres, further cementing itself within Korean society and as a representative of Korean culture.

HOW TO **DRINK SOJU**

How to Order Start with a bottle of soju for every two or three people, deferring to the brand preferences of those at your table if you haven't developed one yet. Each 375 ml bottle (360 ml in Korea) will pour about 6 shots.

You can also order a bottle of beer if you need a chaser or plan to partake in beer cocktails (see Somaek, page 4). Feel free to order a Coke or a Korean cider 사이다, a lemon-lime soda, which can act as a chaser or be used with soju to make different kinds of Poktanju (page 10), or "bomb shots."

Order some anju as well. The clean burn of soju lends itself well to spicy foods, greasy meats, and raw seafood.

HOW TO **OPEN A SOJU BOTTLE**

The Shake Koreans like to shake the bottle before they open it, whether it's a light jiggle or a full tornado (see Soju Tornado, next page). My uncle used to tell me that this was to mix the sugars inside that may have settled in storage, but now I'm realizing that he probably just heard that from Lee Hyori, the longtime spokesperson for Chum-Churum during the mid-2000s. Her TV ads (featuring a little soju tornado!) urged drinkers to shake the bottle to make it smoother. While the jury's out on whether that's true, it's a damn good party trick to herald the commencement of drinking.

The Elbow Tap Using your elbow to tap the underside of the bottle is also thought to mix the soju, but probably harkens back to when soju bottles would come corked rather than capped. Turning the bottle upside down and tapping the bottom a few times dislodged any cork residue, causing it to float to the top. Though the screw cap, introduced in the 1980s, eliminated the need to decork soju, the ritual lives on.

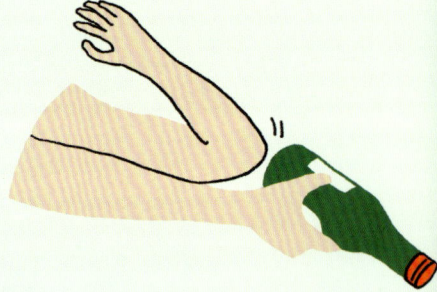

The V-Flick Once you open the bottle, make a V-sign, and hit the neck of the bottle between your pointer and middle finger, so some soju sprays out. You can do the same thing with the blade of your hand, like a light karate chop, or with your elbow. This ritual was also originally used to remove the cork residue but now remains as a fun party trick to put your tablemates in the splash zone. I would probably avoid doing this around elders or businessmen, however.

The X-Open, or the Kim Taehyung A method recently repopularized by the K-pop idol V from BTS, this is extra-expert level. Hold the bottle upside down, wrapping your left hand, thumb-side down, around the cap and your right hand, thumb-side up, around the base (your wrists should make an X). Spin the bottle right-side up while simultaneously cracking it open, ending with your left hand on top lifting off the cap. Add showmanship and flair to impress your drinking buddies even more. To be honest, I'm not sure my academic explanation is doing this method any justice, so just search "How to Open a Bottle of Soju by Kim Taehyung" on YouTube, and best of luck.

Don't forget to save the cap! (See Drinking Games, page 176.)

TURN BOTTLE UPRIGHT

TWIST CAP OFF!

Soju Tornado My cousin Heedae's technique is to hold the bottle in the middle and aggressively shake it like a superfast metronome, rotating his wrist back and forth before stopping abruptly in the middle.

I've found that putting a lateral spin on the bottle will also create the tornado effect, pulling the bottle back toward my shoulder before whipping it forward.

Others swear by the 9–3–12 or 6–3–12 method. Point the cap of the bottle to 9 o'clock, then 3 o'clock, and then 12 o'clock, keeping your wrist as the axle. Now make this motion extremely quickly, snapping to a dead halt upright at 12 o'clock.

Whatever the method, the trick is the abrupt stop, which will reverse the forward motion of the liquid inside the bottle and create the vortex effect. Make sure to keep a solid grip on the bottle—after I taught my friend David the soju tornado, he launched a too-damp bottle clear across a restaurant while trying to re-create it for his friends.

Maekju (Beer)
맥주

The first time I ever had Korean beer was at my father's birthday party, at home in California. It was the only time in my life I've ever really seen him drink. I was maybe thirteen, still too young to drink but old enough to express curiosity, and my dad gave me the go-ahead to have a taste. "Meh." Even from that first sip, I could tell that the beer was watery, flavorless, like slightly skunky bread-water.

The first time I *really* drank Korean beer was about five years later in Seoul, during my final summer before going off to college. We were at a beer bar, known as a hof, post-dinner so that my mom and aunt could continue gabbing while my uncle had a drink (my dad, always a good sport, was just along for the ride). I sat tearing pieces of grilled jwipo (fish jerky) and listening to them talk, until my uncle offered me a glass of beer: "You must be thirsty, might as well have a drink." I took the amber liquid with both hands, bowing slightly, as I'd seen others do. And surprisingly, this time I gulped down the whole glass. It was cold, refreshing, licking the fishy salt off my tongue and leaving me craving more, even as my cheeks warmed from the alcohol.

Beer, or maekju 맥주, as it's called in Korean, is the second most popular alcoholic beverage in South Korea after soju. At first taste, Korean beer can seem bland and boring, compared to the robust malty bocks of Germany, rich velvety brews of Britain, or the sheer diversity of hoppy ales in America. Korean beer specifically has long held a reputation for being, well, just plain bad. But this is by design. Since Korean-style drinking centers heavily around drinking quickly, drinking for a long time, and drinking with food, the specific characteristics of Korean beer are exactly what have implanted it firmly within the country's food culture.

CHEERS: BEER COMES TO KOREA

Historically, most of the beer consumed in Korea has been a very light adjunct rice lager, modeled after an American-style lager that's low in ABV and immensely chuggable.

While there's evidence that Koreans brewed beer during the Joseon dynasty, the beer we know today initially entered the country, like so many things, via Japan. Sapporo was the first beer to be introduced, in the 1870s, followed by Asahi in the 1890s and Kirin in the early 1900s. Even the term *maekju* is of Japanese origin, meaning "barley liquor." In 1933, Kirin established Oriental Brewery Company and Sapporo founded Chosun Beer in Korea. Following the end of the occupation, these breweries were taken over by Koreans—Oriental Brewery is known as OB today, and Chosun Beer evolved into Hite (later HiteJinro).

During both the Japanese occupation and the ensuing Korean War, Koreans were increasingly exposed to Western foods. Beer was first marketed as a foreign-style alcohol to be enjoyed at expensive restaurants, at a time when drinking and dining out was still a new concept for Koreans. But in the 1980s, as the economy bounced back, beer companies began advertising their products to the after-work crowd: perfect for shaking off a hard day with your colleagues. By the end of that decade, beer was, for a time, more popular than soju, especially among office workers and college students.

Somaek: Soju, Meet Beer

Even after soju regained its position as Korea's top alcohol, beer continued to maintain a significant presence. Korean drinking culture is based on getting drunk as fast as possible—and continuing to drink for as long as possible. Korean-style light beer, which focuses less on flavor and more on refreshment and carbonation, is ideal for drinking in large quantities.

Sometimes soju can get you drunk, well, a little *too* quickly. No one wants to be the first to get sloppy, right? But when it's mixed with beer, the resulting combination strikes an ideal balance: easy to chug and just alcoholic enough. **Somaek**, an acronym made with the first syllables of the words **so**ju and **maek**ju, is a popular cocktail among college students and heavy drinkers alike. For this reason, Koreans actively seek out beer that is as light and flavorless as possible, the better to mix with the clear sweet soju.

CHIMAEK 치맥

Where There's Chicken, There's Beer

Sure, your local watering hole might serve up a platter of Buffalo wings with a pitcher of beer. But for Koreans, chicken and beer, known as **chimaek** (a portmanteau of **chi**cken and **maek**ju), is akin to a religion (see page 161).

Chicken first became popular in the 1960s as tongdak, or whole chicken. The restaurant Myeongdong Yeongyang Center introduced electrically grilled rotisserie chicken, with whole birds cooked on revolving spits in an open window convection oven, dripping fat as the skin crisped. (They're still operating today!) Whole chickens became popular among children and families as a cheaper alternative to beef or pork. Then, in the 1970s, tongdak began to be deep-fried, and by the 1980s, KFC had opened its first location in Seoul's Jongno neighborhood, while Pelicana launched the first distinctly *Korean* fried chicken. Fried chicken and yangnyeom (sweet and spicy) chicken become popular fast food, with countless Korean fried chicken shops popping up and peddling half-and-half combos.

But it wasn't until the fervor of the 2002 Korea/Japan World Cup that fried chicken and beer become a true Korean obsession. As the national team climbed into the semifinals on its home turf, spectators across the country gathered in the streets and the pubs to watch and celebrate, chowing down on chicken and guzzling beer to fuel their cheers. The light refreshing style of Korean beer perfectly cut through the greasiness of the fried chicken. The association stuck, and the term *chimaek* was born. Today, chimaek is ubiquitous in Seoul and can be ordered for delivery in less than thirty minutes, becoming a popular choice for al fresco picnics along the Han River.

Chimaek isn't the only portmanteau pairing in the Korean lexicon. Pimaek (pizza with maekju) is popular, while ham-maek (hamburger and beer) is gaining traction as another Western food-and-beer alternative.

TYPES OF **KOREAN BEER**

THE OGs

The first round of South Korean beers, popular in the 1990s, were sweet and light rice adjunct beers that were made to be drunk with food and mixed with soju. You'll still find one or a few of these beers, which are fairly interchangeable in both taste and price, served in cans, bottles, or on draft at nearly every restaurant and pocha.

Hite Introduced in 1993 by the then-named Chosun Beer, Hite is styled after American rice lagers, such as Budweiser. Brewed from barley malt and rice; 4.3% ABV

OB Lager Originally introduced by Oriental Brewery Company in 1952, OB Lager was later renamed just OB in 2003. Made with rice; 4.6% ABV

Cass Originally brewed by Cass Brewery, which was later taken over by Jinro-Coors Brewing Company before being sold to Oriental Brewery in 1999. A pale golden lager made with rice; 4.5% ABV

NEW SCHOOL

In the 2010s, richer, European-style beers hit the market as Koreans' spending and tastes expanded.

Kloud Introduced by Lotte Chilsung in 2014. A European-style malt lager; 5% ABV

Terra Introduced by HiteJinro in 2019. A Czech-style pilsner made with Australian barley; 4.6% ABV

MICROBREWERIES

In recent years, the beer market in Korea has expanded with the increase of foreign beers imported into the country as well as the domestic exploration of brewing other beer varieties.

KABREW A craft brewery that specializes in a wide assortment of beer styles named after significant Korean cities and landmarks (as well as an expanding lineup of canned cocktails).

Magpie Brewing Co. Launched in 2011 with locations in Seoul and Jeju Island, specializing in the kind of nerdy artisanal beer-making you see throughout Brooklyn. It brewed the first sour in Korea and regularly reaches for traditional Korean ingredients like Jeju tangerine and Korean pumpkins.

HOW TO **DRINK KOREAN BEER**

How to Order Beer is most commonly served in bottles at restaurants and pochas. Whenever I order beer for the table in a Korean establishment, I check the size of the bottles on the menu (or eyeball those on the table next to me), because it can be a bit of a crapshoot. Sometimes they're the more standard personal-sized 330 ml (about 11 ounces) and other times they'll come in larger 610 ml bottles (about 20 ounces). At grocery stores, you might find extra-large 1-liter plastic bottles as well.

Bottled beer is often served with maekju glasses that are much smaller than your typical draft beer pint glass, clocking in at 200 ml, about 6.75 ounces, versus a pint glass's 16 ounces. These glasses come in handy for somaek (more on page 4) and shot combos like Milkis (page 20), because they're easy to down in one gulp. Draft beer is usually served in thicker 480 ml (about 16-ounce) steins with handles, also known as hof glasses 호프 잔, that (ideally) have been prechilled or frozen.

Canned beer can be served in 355 ml (12-ounce) or larger 500 ml (16.9-ounce) cans as well, though these are more commonly found at convenience stores, for personal consumption. I like to grab a can for outdoor drinking.

How to Pour I loosely follow the same rules for pouring beer as I do for soju, but much less strictly because the setting for drinking beer is usually much more casual. Plus, the amount of beer a person drinks can be more personal and less dictated by the group. Some of my friends never drink beer because it's too filling, while others will partake only if it's mixed as somaek.

HOW TO **OPEN A BEER BOTTLE**

Sure, you can use the provided bottle opener to pop your beer (there's usually one stored in the chopsticks drawer in the table, or maybe hanging from the BBQ vent above your table). But why not practice perfecting these awe-inspiring alternatives:

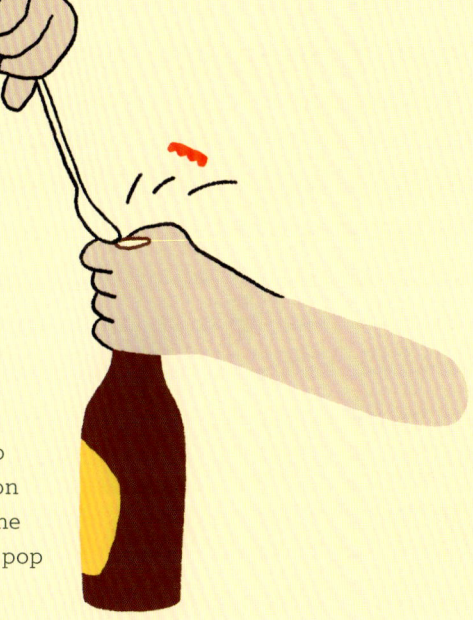

Opening a Beer Bottle with a Spoon Koreans eat with metal spoons and chopsticks, and these utensils are sturdy enough for popping open a beer bottle in a pinch (or for flair). Grip the bottle high up on the neck with one hand, quite close to the cap, and hold your spoon with the other hand. Place the spoon under the ridges of the beer cap. Press down like a lever, using the index finger of the hand holding the bottle neck as a fulcrum, to pop it right off.

Opening a Beer Bottle with Chopsticks Once you've mastered the spoon open, try the same technique using the butt end of a metal chopstick. Got that covered? Take it to god tier by popping open a beer with disposable wooden chopsticks!

Opening a Beer Bottle with Another Beer Bottle
If you've only got beer bottles to spare, use one beer bottle to open another. Steady one bottle on a table and wrap your hand around the neck, just under the cap. Slightly invert a second bottle and align the ridged cap under the edge of the cap of the first. Press down like a lever to open the first beer. You can then use the bottle cap to open the second beer.

WHERE **KOREANS DRINK**

Hof 호프 The hof is a common style of Korean bar or watering hole that focuses on draft beer and Western alcohols. The name is meant to recall German beer halls like Hofbräuhaus and refers to draft beer (although *hof* literally means "courtyard" in German). Hofs also feel a little more like traditional American or European bars: dimly lit, festooned with bowls of bar nuts and TVs playing baseball or soccer.

Convenience Store 편의점 South Korea doesn't have any restrictions on outdoor drinking, which makes the local convenience store a surprisingly hot spot for a tipple. The fridges are packed with frosty soju, cold beer, and even traditional liquors like baekseju (an old-fashioned herbal alcohol), with a cornucopia of anju options ranging from instant ramyun to dried squid. Since most of these mini-marts have seating on premises, it's easy to grab a bottle and a snack for a low-key hang.

포(장마)차

Pojangmacha 포장마차 There's a specific type of romance within the blue tarp walls of the pojangmacha, well documented in many K-dramas and movies. Often abbreviated as pocha, it was originally an outdoor tent-covered stall that specialized in street food and alcohol. The pocha became popular in the 1980s as a cheap, casual place to drink with friends or solo, but started to disappear when the government cracked down on unregulated street stalls during the Seoul Olympics. Nowadays the term includes indoor establishments and is used to more loosely encompass a retro nostalgic vibe evoked by plastic stools, beat-up servingware, and unfussy anju. You can, however, still find outdoor old-school pochas in neighborhoods like Jongno or Euljiro in Seoul.

Modern Cocktail Bar 칵테일바 As Koreans developed more global and expensive drinking preferences, cocktail bars began to pop up in hot neighborhoods like Gangnam and Hongdae (in Seoul) or trendy hotels. These bars are heavily influenced by international cocktail culture and tend to target foreigners and younger imbibers, with craft cocktails featuring imported liquors.

Makgeolli
막걸리

At the base of Namsan Mountain in Itaewon is a makgeolli bar unlike any I'd visited before. In the past I've had makgeolli poured from the same nondescript plastic bottles or copper kettles in hanok-style* restaurants or timeworn establishments, but this spot was dark, industrial, cool. One night, Nick and I tucked into a tiny wooden table near the door to pore over the menu. Various makgeolli and Korean alcohols, most of which I had never heard of before, dotted a matrix arranged from dry to sweet, light to full-bodied. Along with our picks of delightfully fizzy takju† we noshed on unique anju like oysters topped with kimchi kiwi granita and cinnamon garaetteok‡ skewered on campfire-style sticks.

That evening at Mr. Ahn's Craft Makgeolli 안씨막걸리 completely changed my perspective on what makgeolli could be. Makgeolli is an old alcohol, but it's becoming cool again with newer generations. Here, I saw that it could be varied like wine—quirky and surprising. That night also became one of my pivotal life moments, the bar a revolving door of coincidental players: Our friend Doug Park of Tokki Soju stumbled through the front door with Hooni Kim, chef of Meju and Danji in New York, both their faces already ruddy. And at the table next to us sat Mr. Ahn himself, the proprietor of the bar, with Carol Pak, founder of Màkku, an American-born makgeolli company. Carol and I later exchanged information when we realized we were both from New York and I ended up working for Màkku in the year that followed, a crash course on how to redefine an ancient Korean alcohol.

Makgeolli, at first glance, looks so different from any other

* 한옥—a traditional Korean house consisting of an exposed wood beam framework with papered windows and doors

† 탁주—a coarsely filtered version or portion of makgeolli

‡ 가래떡—long, thin Korean rice cakes

alcohol. It is creamy white and milky, with light lactic bubbles popping around the edges. Since the liquid is not fully filtered, rice sediment often settles at the bottom of the bottle, which can be mixed back in before serving. It's classically drunk out of shallow copper-aluminum bowls that recall rice bowls, and its flavor can range from sweet and yogurty to savory and tart.

In America, people sometimes refer to it as a rice wine or a rice beer, but there's nothing else that's really like it. Makgeolli is iconically Korean, even more than soju. Not only is it the country's original alcohol, first brewed around the first century BC, its short, straightforward fermentation means it can be easily produced in homes or in small batches. That's why there are endless varieties of makgeolli, including many regional versions produced across the country with differing flavors and tastes, much like that of natural wine. This makgeolli reflects the unique terroir and personality of Korea's landscape.

But unlike wine, makgeolli is difficult to store for long periods of time and too volatile to ship long distances in its natural state—like kombucha, it keeps fermenting. Commercial makgeolli exported from Korea is usually made with added preservatives and pasteurized, which can mute or flatten the alcohol's subtle flavors. But a growing market and demand for this fizzy white alcohol is giving rise to local producers and home brewers in the United States.

AN ALCOHOL BY MANY DIFFERENT NAMES

Makgeolli goes by several names and can be found in a few different forms.

Makgeolli 막걸리 ("roughly strained")

Refers to the drink as a whole

Nongju 농주 ("farmer's wine")

Another name for makgeolli, referencing its humble origins

Takju 탁주 ("opaque wine" or "turbid liquor")

Yet another name for makgeolli, this term also specifically refers to its cloudy mixed form, as well as the milky part of the makgeolli when it separates.

Cheongju 청주 ("clear wine")

The clear rice wine that's filtered from makgeolli when the takju is separated out. Cheongju has commonly been used in traditional rituals and rites, as well as for cooking.

It's sold separately from makgeolli as well: Lotte has been making a popular commercial version of cheongju called Chungha 청하 since 1986, while Yangchon Brewery brews a cheongju made with rice grown using snails in lieu of herbicides, called Yangchon Snail Rice Cheongju 양촌 우렁이 쌀 청주, which is found at many Michelin-starred Korean restaurants in America today.

Yakju 약주 ("medicinal wine")

Another name for clear cheongju

Beopju 법주 ("law liquor")

Yet another name for cheongju, a traditional liquor that dates back to the city of Gyeongju during the Goryeo era (918–1392)

Danyangju 단양주

When rice, nuruk (fermentation starter), and water are brewed together once, this is called danyangju. Makgeolli is mostly made from danyangju that has been filtered.

When more rice and nuruk is added and brewed again, it becomes iyangju 이양주, and then samyangju 삼양주 when brewed for a third time.

THE MOST ANCIENT OF ALCOHOLS

Before there was soju, there was makgeolli. Records of its production date back to before the Three Kingdoms period (57 BC–668 AD). During the Goryeo era, an early form of makgeolli called ihwaju 이화주 was made during pear blossom season by mixing rice, nuruk (fermentation starter), and dduk (Korean rice cake) with a little water to make a thick spoonable alcohol that was served in the royal courts. Since it is so easy to brew at home, many farmers made makgeolli with their leftover rice harvests, and it grew to become a popular layman's drink. (It had the added benefit of being nutritious and restorative, and could be drunk as a meal substitute in a pinch.)

Although home brewing was banned during the Japanese occupation, makgeolli remained popular until the 1960s, when the use of rice in making alcohol was prohibited. Unlike the industrialization of soju, however, the efforts to industrialize makgeolli were unsuccessful—brewing this home-style liquor on a large scale proved too difficult, and using cereal grains like wheat or corn couldn't replicate the same flavor. By the time rice was reintroduced into the makgeolli-making process in 1977, makgeolli had already gained a reputation for being low-quality, and most people had moved on to imported alcohol (like beer), a preference that lingered until an artisanal resurgence in recent years.

MAKGEOLLI TODAY

During a trip through the southern cities of South Korea, I tasted a delicious craft makgeolli called Boksoondoga 복순도가 and purchased a few bottles to give to my family as gifts. It's known as the champagne of makgeolli because the bottles can be extremely volatile and overflow when you open them (they come with precise instructions on how to best tackle the task). I handled mine with great care on the train ride back, but a week later a bottle exploded in my aunt's kitchen when she, forgetting the warning, tried to open it.

I think that fermentation is the defining foundational technique of Korean food, which is why doenjang and gochujang are such key components of our cuisine. Makgeolli has a quicker fermentation and production process than most alcohol, which makes it easy to play and experiment with. Many different provinces in South Korea produce makgeolli, usually served in countryside restaurants, and there are more than seven hundred small-scale makgeolli breweries today. Each makgeolli tastes different, based on the local rice used or a unique technique. For example, on Jeju Island they serve a peanut makgeolli that features peanuts grown on Udo Island. Artisanal makgeolli has also boomed in recent years as younger generations revisit and play with ancient techniques, as have makgeolli bars, like Mr. Ahn's Craft Makgeolli, where you can try these different types of makgeolli and other traditional Korean alcohols.

The natural fermentation of makgeolli means it must be consumed fresh within a few days, so it's hard to ship overseas. (Wait too long and the makgeolli may turn sour, or explode, like in my aunt's kitchen!) The Korean makgeolli companies that do export to America generally pasteurize their product or add preservatives. Both methods tend to alter the taste, and most of what's available for purchase in America is not nearly representative of the vast diversity of makgeolli back in Korea.

But thankfully, there is modern makgeolli being brewed now in America as well. Alice Jun's Hana Makgeolli opened a brewery in Greenpoint, Brooklyn, in 2020 and specializes in traditional makgeolli like hwaju, yakju, and takju, and Carol Pak's canned Màkku comes in flavored versions like blueberry and passion fruit in addition to classic rice. The range in market positioning of these two brands reflects the diversity of makgeolli: Hana Makgeolli is distributed as a natural rice wine with a higher price point that's meant to be paired with dishes at restaurants, while Màkku is marketed as a more accessible rice beer that's carried in bodegas and Asian grocery stores.

SO EASY YOU CAN MAKE IT AT HOME

Makgeolli has been brewed in Korea for thousands of years because at its core it's a simple fermentation, something that can be made in your kitchen with just rice, water, and nuruk (fermentation starter). It really is that easy—I have made makgeolli at home with ingredients sourced from my local H Mart. Just steam some rice (usually glutinous rice), make a porridge with nuruk and water, and let it ferment in a hangari 항아리 (a Korean earthenware pot) for a week or two. Once strained, the resulting liquid is **takju**, and the clear liquid that further separates from the takju is known as **cheongju**.

MAKGEOLLI

CHEONGJU

TAKJU

HOW TO **BUY**

Makgeolli is mostly sold in 750 ml plastic bottles that are usually green or white. The commercial versions, especially those exported to America, are pasteurized to prolong shelf life, but most local varieties are not. It's here that makgeolli's unique charms really shine. Travel to any province in South Korea and they'll have a local variety of makgeolli.

HOW TO **OPEN**

You have two options when drinking makgeolli:

Shake Sediment naturally settles at the bottom of the bottle, so gently shake before opening. You can also tap the bottom gently to dislodge the sediment. Swirl the bottle, holding the bottle upside down or upright—no need to shake like a piston. Open the cap carefully, being prepared to close it up again to prevent overflow.

Don't Shake Some Koreans prefer to not shake up the sediment and just drink the clear liquor, aka cheongju, because they believe that this will lessen the following day's hangover. Simply open the bottle and pour, stopping before you reach the bottom sediment.

HOW TO **SERVE**

As a nod to its humble roots, makgeolli is traditionally (but not always) decanted into teakettles and served in aluminum/copper rice bowls (sometimes with a handle). Drinking out of these wider serving vessels rather than glasses allows the scent of the makgeolli to really open up, and since makgeolli has a lower ABV, it can be drunk in larger quantities.

WHEN TO **DRINK**

On a Rainy Day When it rains, the soft pitter-patter of drops hitting the ground conjures up cravings for jeon and makgeolli for Koreans, because it mimics the sound of the sizzle of the fritters frying in the oil (so romantic, right?).

On Top of a Mountain Koreans love to drink makgeolli when hiking—they'll carry a bottle to the top of a mountain to crack open as a well-deserved refreshment, or drink at a mountainside restaurant with classic makgeolli anju, such as acorn jelly salad and buckwheat pancakes.

For Ancestral Rites or Chuseok Makgeolli is commonly served during funeral or ancestral rites, as well as during harvest festival celebrations like Chuseok.

Clockwise from far left:
Bokbunja-ju (page lxix),
Beopju (page lxiii),
Maesil-ju (page lxx),
Sansachun (page lxx),
Cheongju (page lxiii),
and Baekseju (page lxix)

Sool

술

The spectrum of Korean alcohol, of course, extends much further than just soju, beer, and makgeolli. There are so many others with storied histories not commonly found outside of Korea. Here's a short list. Look for these at your local Korean grocery store or the next time you find yourself traveling in Korea.

Baekseju 백세주 ("hundred-year wine") An alcohol fermented with various herbs, including ginseng. It's named as such because it's thought to make the drinker live to one hundred years old. It can easily be purchased at Korean convenience stores.

Beolddeokju 벌떡주 ("wake-up alcohol") I love asking my mom about this herbal rice wine because it makes her giggle every time—the term literally means to "stand up" or "become erect." The drink is infused with ingredients that are believed to increase male stamina, and, appropriately, the bottles are topped with a ceramic penis with a grinning face. You can find this medicinal but fruity alcohol often sold as souvenirs at gift shops and regional markets.

Bokbunja-ju 복분자주 (black raspberry wine)
A fermented fruit wine made with bokbunja 복분자, or black raspberries, ringing in at about 12% to 20% ABV. Sweet and syrupy, it originates largely from Gochang in Jeonbuk-do, where bokbunja grow wild in the mountains. In the 1960s, those living near Seonunsan 선운산 began actively growing bokbunja in their fields, and in 1994 the first commercial producer of bokbunja-ju, Gochang Specialty Bokbunja-ju 고창명산품복분자주, was established there. Today, Bohae 보해양조, launched in 2004, is one of the larger makers of the bokbunja-ju commonly found stateside.

Dongdongju 동동주 ("floating alcohol") A type of cheongju that's often served with a few floating grains of rice, hence the name.

Dugyeon-ju 두견주 (azalea wine) A traditional rice wine brewed with dried azalea flower.

Gamju 감주 Also known as dansul 단술, a takju-style liquor made with rice and yeast that's similar to sikhye, a sweet nonalcoholic rice-based drink.

Insam-ju 인삼주 (ginseng wine) A medicinal wine made from ginseng, often by infusing whole pieces of ginseng root in a high-proof soju. Due to the rarity of wild ginseng, insam-ju is quite expensive and highly prized.

Maesil-ju 매실주 (plum liquor) An alcohol infusion made with ripe or green plums that dates back to the Goryeo dynasty. See page 42 for a recipe.

Moju 모주 A Jeonju specialty beverage made by boiling makgeolli with ginger, jujube, licorice, arrowroot, cinnamon, herbs, and spices. It contains less than 1% alcohol.

Munbae-ju 문배주 ("wild pear alcohol")
A traditionally distilled liquor historically popular among kings. Its name comes from its fruity scent, even though the liquor doesn't actually contain any pear.

Sansachun 산사춘 A light, sweet drink with a slight rose color. It's made with sansuyu 산수유, a small red berry, and tastes similar to a semisweet white wine.

Omegisul 오메기술 A traditional liquor brewed with omegitteok 오메기떡, a steamed millet cake that's a signature of Jeju Island.

Sogokju 소곡주 A traditional scholar's liquor from the Baekje dynasty (18 BC–660 AD), also known as anjeun-baengi sul 앉은뱅이 술, aka the "sit-down drink," because it's purportedly so good it prevented scholars from standing up to go take their exams. It's fermented with glutinous rice and yeast.

HA
1차

EASY DRINKING

KOREAN
SHOTS,
DRINKS &
INFUSIONS

You've ordered your first bottle of soju, poured your first shot, and tasted the first sip. Sigh, maybe it's not really for you. Now what?

While soju is deeply interwoven into the fabric of Korean drinking, there are, of course, many, many people who are not interested in drinking it straight. I get it—either it doesn't taste like anything or it tastes too much like a straight alcohol burn, which can be hard to get used to. That's why there are numerous quick and easy soju hacks in Korea, which are popular especially if you're just beginning to explore soju.

We have to start with **somaek,** a combo of soju and beer that seems deceptively straightforward until you uncover countless ways to measure, execute, and imbibe this two-for-one. In this chapter, we'll also drop some **poktanju,** or bomb shots, like the sweet and frothy Milkis Shot (page 20) and stir up **quick drinks** that take less time to make than to crack open another bottle of soju. After all that easy swigging, we'll settle down with **soju infusions,** or damgeum-ju—simple ways to make your own flavored soju.

Most of these recipes are short and accessible, easy to deploy with ingredients you probably already have at home or can find out at the bar. Get a group of friends together to try them all, and see what you like best.

Somaek (Soju & Beer)

소맥

Somaek is kind of the national cocktail of Korea, its name a portmanteau of its two ingredients: soju and maekju (beer). But it's not ordered at the bar—it's a cocktail you make yourself or for one another. The soju should be of the green-bottle variety, and the beer should be light, like Hite or Terra, but you can also use any light lager-style beer, such as Budweiser or Coors. But from there the proportion and execution possibilities are endless.

I particularly like somaek when drinking with friends. It's a quick way to down the first few bottles and do away with the rituals of pouring, which can get a bit tedious as the night wears on. It can also be hugely entertaining (and dangerous!) trying to one-up each other by executing the perfect somaek spoon slam or downing a glass fastest.

Somaek is often served in its own special glass, a smaller 225 ml (8-ounce) Korean-style beer glass. Some somaek glasses have printed markings along the side indicating different somaek ratios. You could spend an entire night making, drinking, and arguing over the perfect ratios and recipes, only to wake up the next morning having to start anew. But in the meantime here are a few jumping-off points—use a jigger or a soju shot glass as a measure, or just eyeball it.

COMMON SOMAEK PROPORTIONS

	RATIO OF SOJU: BEER	OUNCES	SHOTS
BABY STEPS	1:10	0.5 ounces soju 5.5 ounces beer	¼ shot of soju 2¾ shots of beer
SMOOTH	1:5	1 ounce soju 5 ounces beer	½ shot of soju 2½ shots of beer
STRONG	1:2	2 ounces soju 4 ounces beer	1 shot of soju 2 shots of beer
DEADLY	1:1	3 ounces soju 3 ounces beer	1½ shots of soju 1½ shots of beer

WAYS TO **MIX SOMAEK**

Koreans believe that creating foam (which they call the "cream") when mixing a somaek is important because it shows that the drink is well combined. Creating it is purported to release the gas that would otherwise make you feel very full and uncomfortable. Here are a few techniques:

Swirling Swirl the glass to gently stir the somaek together and release bubbles.

Spoon slam Slam a spoon into the glass, causing the somaek to mix and foam.

CLINK

Chopstick tapping Insert one chopstick in the drink and use the other chopstick to tap aggressively.

Spoon clap or cappuccino The head of a jang (fermented sauce) company in Pohang introduced this one to me. Place two spoons back to back in the glass, and clap them together to mix. It's called a cappuccino because of the resulting foam, but I also like to call it a clap-puccino.

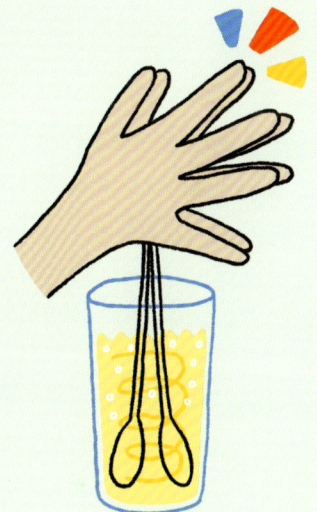

Napkin Cover the glass with a napkin, cover with one hand and slam on your other palm.

Extra Credit—The Shower Ju Feeling confident with all of the above? Get extra points by making a beer shower! Pop the top off a bottle of beer and cover the opening with your thumb. Shake vigorously, then move your thumb slightly to release a tiny opening so that the beer begins to spray out. Spray into beer glasses prepped with shots of soju, into your own mouth, or all over your friends!

SHOTS

Poktanju (Bomb Shot)
폭탄주

Sometimes soju feels too strong to drink straight, but beer feels a little too light and watery to really get you going. Enter the poktanju, or bomb shot. Also known as a soju bomb, it's the most exciting way to drink somaek (in my opinion). Instead of mixing soju and beer in a glass, just drop a shot of soju into a glass of beer and chug!

WHY WE BOMB

Koreans like to be their own performance mixologists, combining shots or bottles of soju, beer, makgeolli, and other liquors to make poktanju. It's a fast, fun way to pop some bottles, get drunk, and play with your fellow drinkers. Since the standard soju is flavorless and tastes intense to novice drinkers, mixed drinks that are consumed quickly like this have become a traditional introduction to drinking for many Koreans.

Poktanju in Korea emerged initially as a feature of military drinking culture,[*] possibly originally introduced by US soldiers in the 1960s and '70s. It blew up in popular culture following the National Defense Committee incident 국회 국방위 회식 사건 in March of 1986, when a meeting of high-level parliament and military generals broke down into a poktanju-fueled fistfight. Representative Nam Jae-hui threw a glass, a shard of which lodged in Major General Lee Dae-hui's eyelid, and he in turn kicked Representative Nam. The whole situation resulted in a PR scandal that was covered widely in the media, making poktanju popular among the public.

But it's not just soju bombs. Rack up your glasses, chill your beverages, and get ready to power through these explosive combos.

[*] In South Korea, all men of age must serve in the military.

My friend Mickey introduced us to this somaek-in-shot-glass format one night after a kimjang (kimchi-making) party at my house, and we've never gone through bottles of soju faster. I'm not sure how much of it is a placebo effect that tricks your brain into thinking it's drinking beer when—surprise!—it's soju. I've also heard of re-creating this in a larger format using a Korean beer glass (as in, fill the glass with soju and top with a tiny splash of beer), but proceed with caution before attempting. This will get you drunk fast.

Fill a 2-ounce soju glass (or a regular shot glass) 90 percent full with soju. Top with beer and shoot immediately. The longer the shot sits, the less effective this technique, as the beer will start to mix with the soju.

KKULJU (HONEY SHOT)
꿀주

Soju

Light beer, such as Terra or Budweiser

The first somaek I learned to make was actually a line of sake bombs. In college, we'd wreak havoc on our local Japanese restaurant, lining up rows of beer and shot glasses balanced on chopsticks, slamming our fists on the table to drop the liquor and get our night started. The Seoul Train (or Somaek Domino) is assembled and discharged in the same manner. I know it's a horribly disruptive scene, but it still thrills me to see this executed in the right context.

I'm particularly in awe of the epic Seoul Train executed in Psy's "Hangover" music video, which is my forever inspiration. In real time, Psy gently taps a single shot glass and 100-some shot glasses topple in around him as he dances away.

You'll need one shot glass and one beer glass per person.

SEOUL TRAIN
소맥 도미노

Light beer, such as Cass or Coors

Soju

Line up in a row as many beer glasses as there are people and fill each glass halfway with beer. Balance the shot glasses on the rims in between the beer glasses (you should have one leftover) and fill them with soju. Use the last shot glass to knock one of the end shot glasses into its beer glass, creating a domino effect. Chug the resulting somaek quickly.

ALTERNATIVE EXECUTIONS

• Use a metal chopstick to poke a hole into the top of a beer cap. Place your thumb over the hole, shake the bottle and quickly turn it upside down to spray beer into the beer glasses.

• Slam your fists (or your head) onto the table to trigger the shot glasses to fall in.

• Use a beer bottle like a golf club to knock the soju glasses in.

MILKIS SHOT
밀키스주

½ shot soju

½ shot lemon-lime soda, such as Chilsung Cider or Sprite

½ shot light beer, such as Terra or Cass

Milkis, a yogurt-flavored carbonated soda, was the holy grail drink of my childhood. The slim white can beckoned me toward the vending machines at subway stops and underground shopping malls in Seoul, and I'd endeavor to find the right moment to implore my mother for 100 won so I could buy one.

The Milkis shot, however, contains no Milkis. Instead, it's a once-popular shot combo in Korea that was introduced to me by my friend Jihan on a particularly random and raucous night out at Gopchang Story BBQ in Manhattan's K-town. The simple ratio of ingredients, to which great force is applied, results in a creamy concoction that tastes, surprisingly, like my favorite Milkis!

Add all the ingredients to a small glass; a somaek glass is ideal. Cover the glass with a stack of paper napkins and seal with the palm of your dominant hand. Forcefully smack the glass down against the heel of your other palm and then thrust forward to mix. This should cause the mixture to cream and foam together. The recipient of the glass should down it in one shot for the full effect.

ALTERNATIVE EXECUTION

- You can also mix the ingredients together in a glass and drink (skipping the smacking); the resulting drink will still be tasty, though less creamy.

The name of this cocktail translates to "after hard times comes good times," a reference not only to when you should drink it (after a long hard day), but also the drink itself: Meant to be downed in one shot, it starts strong and bitter as the soju first hits your throat but ends sweet, with the pleasant fizz of Coke on your tongue.

GOJIN-GAMNAE-JU
고진감래주

Coke

Soju

Light beer, such as Budweiser or Terra

1. For each drink, prepare two 2-ounce soju glasses and a Korean-style beer glass. (You can also build with two regular shot glasses that stack into each other and fit snugly but not too tightly into a small glass, though the result may vary depending on the size and shape.)

2. Fill one soju glass halfway with the Coke and place it into the bottom of the beer glass. Nestle the second soju glass on top of the first and fill with soju. Pour the beer carefully into and around the empty space to top.

3. Drink the whole glass in one go, without mixing. Rinse and repeat for your friends.

A LOVE LETTER TO
THE POHANG POKTANJU AJUMMA

When we visited my friend James (known on the Internet as Jamesyworld) and his family in his hometown of Pohang, he invited us to visit the famous "Poktanju Ajumma," who happens to be friends with his mom. She owns a pub on a beach-lined street downtown, where she became famous for her skill at making poktanju (see page 10), leading to multiple appearances on Korean variety shows. When we visited, she rolled out a cart laden with beers and soju and began her performance. First, she popped the tops off the beers using the soju bottles themselves (I still don't understand how). After pouring soju into our glasses, she shook up a bottle of beer and sprayed it first into the glasses, then into our mouths, and then more into the glasses. As the beer foamed like whipped cream, she passed out the glasses, which she firmly instructed that we down in one gulp. The result? Smooth and sweet as a meringue. I'm pretty sure she proceeded to dazzle us with various other poktanju, but nobody recalls anything else that happened that night . . .

EASY DRINKS

While soju is most often drunk straight and by the shot, these simple mixed drinks are uncomplicated and easy cocktails that are even easier to imbibe. They require no extra prep or subrecipes (hop over to 3-Cha, page 102, when you're ready to break out your barware).

Simple, hydrating, and effective (at getting you drunk). You've probably had a gin and tonic or a vodka tonic before—the concept is similar here, but the amount of soju can be far greater than that of other liquors due to its lower alcohol content. I like a 1:1 ratio; start with these proportions and adjust according to your preference.

In an ice-filled glass, stir together the soju and tonic. Squeeze in the lemon, which you can throw into the drink as well.

SOJU & TONIC
소주 토닉

Makes 1 cocktail

Ice

3 ounces soju, such as Chamisul

3 ounces tonic water

Lemon wedge

SOJU HIGHBALL
소주 하이볼

Makes 1 cocktail

Ice

1½ ounces soju, preferably
Tokki Black or Tokki Gold

4 ounces club soda

Lemon peel, for garnish

The highball is the perfect vehicle for enjoying your favorite artisanal soju, especially one that's higher-proof. Highballs may seem basic, but basic works when it's efficient and delicious at the same time. This cocktail is akin to a vodka soda, the daytime drink of choice for most of my hard-drinking American friends. It's clean and fizzy with just a tiny hit of lemon.

Fill a collins glass with ice. Add the soju and club soda, and stir gently to combine. Squeeze the lemon peel lightly over the drink to express its oils, rub along the rim, and add to the drink.

TOKKI SOJU

Tokki Soju Black Label is 80 proof (40% ABV) versus its White counterpart, which is 23% ABV. Brandon "Bran" Hill, master distiller of Tokki Soju, specifically developed Tokki Black for use in cocktails, which makes it a delicious base for the Soju Highball.

Tokki Soju Gold Label, which is aged in oak barrels like whiskey and comes in at a slightly more potent 46% ABV, makes for an even more interesting take to rival the popular Japanese whiskey highball.

I've been a massive fan of this Brazilian cocktail ever since a trip to Rio de Janeiro, where I had numerous versions over the course of many nights. Traditionally, the drink is made with cachaça, a clear Brazilian spirit distilled from sugarcane that's not dissimilar to soju (though much more potent). The sugar and lime mix so seamlessly with the liquor that you won't see the alcoholic punch coming until you've had three.

Add the sugar and the lime slices to a glass and muddle using a muddler or the handle of a wooden spoon until the lime is crushed and the sugar dissolves. Add the soju and fill the glass with ice, preferably crushed. Stir, sip, and enjoy.

SOJU CAIPIRINHA
소주 카이피리냐

Makes 1 cocktail

1 tablespoon sugar

½ lime, sliced

4 ounces soju of any kind

Ice, preferably crushed

MAKSOSA (MAKGEOLLI-SOJU-CIDER)
막소사

I love the name of this cocktail for two reasons: First, it follows the very Korean way of naming any combination of things by just taking the first syllable of each ingredient, and, second, the resulting word salad sounds like mapsosa 맙소사, *an exclamation akin to "Oh my goodness!" Which is what you'll be saying after trying this surprisingly delicious drink, equal parts fizzy, boozy, and sweet.*

Makes 1 cocktail

2 ounces makgeolli

2 ounces soju

2 ounces lemon-lime soda, such as Sprite or Chilsung Cider

In a glass, stir together the makgeolli, soju, and lemon-lime soda. (Or make a large batch, with equal amounts of all, in a pitcher or a kettle to share.)

KEY TONIC →
키토닉

This is a soju cocktail created and popularized by Kim Ki-bum, aka the singer Key from the group SHINee, a second-generation K-pop boy band. On the show I Live Alone, *he makes this cocktail—a mix of steeped black tea and soju—for his fellow ex-military friends. Key's invention (which he dubbed "Ki-bum-ju" 기범주) went so viral that he later teamed up with the company Jinro to make Key Tonic, a nonalcoholic mixer that's made to be combined with a bottle of soju.*

Serves 4

2 cups (or 550 ml, per Key's original recipe) brewed black tea, such as Earl Grey

1 bottle soju, such as Jinro Is Back

In a pitcher or teapot, mix the brewed tea and soju, adding ice to taste. Serve while bopping to SHINee's hit songs "View" or "Ring Ding Dong."

Bokbunja-ju, a black raspberry wine, is one of those drinks I sometimes order because it sounds good and looks beautiful, but then I realize—oh no—it's just way too sweet.

Until, that is, our friend Wonny, co-owner of the NYC-based Korean perfumery and café Elorea, shared this elegant solution one night at dinner in K-town. He poured equal parts bokbunja-ju and our favorite soju into a soju glass, and the resulting combination was so pitch-perfect that we were clamoring for more. Now this is the welcome drink we make to kick off every group dinner outing together.

Simultaneously pour the bokbunja-ju and the soju into a pitcher. Serve in soju glasses, cheers-ing endlessly over copious amounts of food.

WONNY'S BOK-SO-JU
복소주

Serves 8

1 (375 ml) bottle bokbunja-ju (Korean black raspberry wine)

1 (375 ml) bottle soju, such as Jinro Is Back

INFUSIONS

Hometown Ju-Ju-Ju

When I imagine the dream house I will one day retire in, it looks a lot like the renovated hanok* that Chief Hong lives in on the Korean drama *Hometown Cha-Cha-Cha* 갯마을 차차차: It has paper-covered lights that hang from the ceiling, with vintage radios and old books lining the wooden shelves. But my favorite feature is the divider between his living space and his kitchen, adorned with large glass jars of liquor. These jars are Hong's collection of damgeum-ju, or infused liquors.

Damgeum-ju 담금주, also known as yaksul 약술 (medicinal liquor), refers to any type of infused spirit, but is most often made with soju. You might see them at your harabeoji's† house or an old-school restaurant: long thin jars with gold lids, spindly tentacles of ginseng or bellflower root tracing their walls. Damgeum-ju dates as far back as soju itself, when during the Goryeo dynasty distilled soju was often mixed with medicinal ingredients to combat the damage caused by excessive drinking. Ever since, it's been a fantastic way to preserve roots, fruits, and even flowers, which slowly impart their fragrance and flavor to the alcohol they are submerged in.

The best part about damgeum-ju? They're super easy to make and play around with at home, as long as you have space on your shelf and time on your side. Supermarkets in Korea sell special 30% ABV soju (which is also called damgeum-ju) that's made for infusing. This can be a bit harder to find here in the States, so I've developed these recipes to mostly use 20% ABV soju, which comes in the more easily accessible 375 ml glass bottle or larger 1.75-liter plastic PET bottle. Feel free to use any (unflavored) soju you have on hand: Lower ABV ones might extract a little less flavor, but you can infuse them for a longer time to make up for it.

---

* 한옥—a traditional Korean-style house with roughly hewn wooden beams and paper walls

† 할아버지, grandfather

Try these recipes to start, and then experiment with your own! Label each bottle with the date and weight or volume of ingredients so you can reference or replicate in the future. Once they're ready, I like to drink them straight to really taste the subtle notes of the infusion, but you can also add tonic or soda for an easy spritz (see the Soju Spritz on page 109) or experiment with using your damgeum-ju in some of the other cocktails in this book.

MAESIL-JU
매실주

Makes 2 quarts

3 pounds green plums (labeled as maesil or ume at Asian grocery stores)

1 (1.75-liter) PET bottle soju (at least 20% ABV), such as Chamisul Original or Jinro 24

½ cup vodka

Maesil (green plums) pop up at my local H Mart right when spring starts to awaken the trees in my neighborhood. They're not really that good to eat as is, but they're perfect for making pickles, syrup, or alcohol to capture their sweet spring flavor. Maesil-ju is one of the more classic (and approachable!) soju infusions, since the bouncy green plums are available fresh for only a short time but last forever when preserved.

Here, I mix some vodka in with the soju to help with the extraction—you'll need the alcohol to be at least 30% ABV to extract enough flavor from the hardy plums. With this—and any—fruit-based damgeun-ju recipe, use good, unblemished fruit, and infuse for no more than 100 days; otherwise, any seeds may start to react to the alcohol. Most maesil-ju recipes incorporate sugar or honey, but I skip it because I prefer a tarter flavor and the naturally subtle sweetness that occurs during the infusion. If you prefer it a little sweeter, you can stir in some brown sugar after you strain out the fruit.

1. Wash the green plums thoroughly and use a chopstick or toothpick to remove the stems. Let them dry in a sunny, airy place for one day.

2. Clean and sterilize a ¾-gallon or 1-gallon jar with a lid (there are many fermentation-grade crocks you can use, but I just use glass jars with tight-fitting silicone-rimmed lids from IKEA). Add the plums, soju, and vodka. Store, covered, for about 100 days, stirring once a month or so, then strain out the fruit.

3. You can enjoy the alcohol immediately, or let it ferment for another 3 months for a deeper flavor.

DON'T TOSS THE PLUMS!

I like to save the drained green plums to make my own version of maesil-cheong, a Korean fermented syrup. Weigh the fruit and mix with an equal weight of granulated sugar. Store in a sterilized jar or fermentation crock for 100 days, mixing occasionally, before straining out the fruit. The result has a deeper level of flavor compared to the supermarket variety. Use in cocktails like the Maesil Old-Fashioned (page 113), mix with some seltzer or club soda for an easy nonalcoholic spritz, or use as a sweetener when cooking (or in your maesil-ju!).

In the '90s, "cocktail soju"—a sweet mix of soju, fruit juices (like cherry, lemon, orange), and soda—became momentarily popular at soju bang, café-like hangout spots for couples and friend groups to drink. This inspired larger soju companies at the time to make their own bottled versions, predecessors to the fruit-flavored soju that popped up in the mid-2010s and remain popular today.

This berry ju is an easy infusion that relies on the fermentation of natural fruit to create a fruit soju. The result tastes like what you wish that chemically processed fruit soju would taste like: juicy, naturally sweet, and delightful. Try this with a mix of strawberries and blueberries or blackberries, as below, or experiment with your own fruit mix. All you need is a 2:1 ratio (by weight) of fruit to soju, and make sure that the soju or alcohol you use is at least 30% ABV; otherwise, the fruit may spoil.

1. Sterilize a ½-gallon jar with a tight-fitting lid. You can do this by washing it with dish soap, drying it completely, then adding boiling water and putting the lid on to let the steam disinfect it, or by immersing the whole jar in boiling water for several minutes.

2. Wash and dry the berries and slice the strawberries, if using. Transfer the berries to the clean jar. Add the sugar and mix everything to combine. Put the lid on and let the jar sit for 2 to 3 days, stirring once a day.

3. Add the soju and vodka and mix to combine. Put the lid on and let the jar sit for 100 days. Then strain into a bowl and return the liquid to the jar, discarding any fruit solids. Ferment the liquid, with the lid on the jar, in a cool, dry place for another month before drinking.

BERRY JU
베리주

Makes 1 quart

1 pound (500 g) berries
(a mix of strawberries and
blueberries or blackberries)

½ cup (100 g) sugar

500 ml (1 pt) soju (at least 20% ABV),
such as Chamisul Original or Jinro 24

500 ml (1 pt) vodka (40% ABV)

JUJUBE JU
대추주

Makes 1½ cups

¼ cup dried jujubes
1 (375 ml) bottle soju (20% ABV)

For me, dried jujubes, also known as red dates or Chinese dates, feel like a cornerstone of Korean culture. When I was a kid, sliced rounds adorned the creamy white slabs of steamed dduk (rice cakes) that my halmeoni would buy from the local rice-cake shop (I always picked them off). When I got married, our parents tossed handfuls into my hanbok skirt to predict the number of children we might have (22?!). And every Chuseok (the harvest holiday colloquially known as Korean Thanksgiving) I mound a tall pile of jujubes as one of the table centerpieces, crawling under the table to retrieve any runaways.

Jujubes are used as a base flavor in so many core Korean foods. They can be steeped in tea, infused into rich stews like samgyetang, and even dropped into the salt brine when making traditional doenjang (fermented soybean paste) and ganjang (Korean soy sauce). So infusing this rich red dried fruit into soju gives you a drink that feels nostalgically Korean and tastes like a nutty sherry. The longer the infusion sits, the more syrupy and sweet the liquor becomes. Drink it on its own or in the Jujube Ginseng Negroni (page 117).

1. Cut the jujubes in half to expose the insides.

2. Combine with the soju in an airtight container and infuse for at least 1 day before drinking. You can infuse for longer if you'd like—the liquor will get sweeter with time.

Samgyetang is a cozy ginseng chicken soup in which the cavity of a young hen is stuffed with glutinous rice and a fistful of Korean herbs. Weirdly, it's highly sought-after in the summer, even though a bubbling hot cauldron seems like the last thing you'd want on a 90-degree day. The restorative dish is supposed to make you sweat out toxins, replacing them with the healing properties of the ginseng and medicinal herb–infused chicken and broth. I love it year-round because, well, what's not to love about chicken soup?

Most Korean supermarkets sell these samgyetang herbs as a single-serving meal kit, so it's an easy way to make a small portion of damgeun-ju. Drink this ju—which smells a little like a homey medicinal shop and tastes both savory and spiced—by the glass or like we do at Orion Bar, paired with a bottle of Bacchus-D, that Korean old-man energy drink, for an invigorating one-two punch.

In a wide-mouthed jar or glass bottle with a tight-fitting lid, combine all the ingredients. Infuse for at least 3 days, or longer for a deeper flavor.

SAMGYETANG-JU
삼계탕주

Makes 1½ cups

1 (25 to 30 g) serving pack samgyetang herbs, or an equivalent mix by weight of any or all of the following: dried ginseng, jujube, chestnut, licorice root, cinnamon, milk vetch, mulberry bark, and angelica root

1 (375 ml) bottle soju (20% ABV)

SOJU APERITIF
소주 아페리티프

Makes 1½ cups

1 (375 ml) bottle soju (20% ABV),
such as Chamisul Original

2 tablespoons dried omija

Four 2-inch strips of grapefruit peels

1 tablespoon dried mugwort

In Italian culture, aperitifs like Campari, Fernet, and Aperol are drunk before a meal to whet your appetite, though I'll also reach for them after to help settle my stomach. I wanted to make a Korean aperitif with similar fruity and bitter ingredients that was inspired by the many medicinal alcohols in Korean history. Soju is lower in alcohol percentage than most aperitifs, so you'll get a subtle extraction that will intensify the longer you let it steep. Strain and sip straight after dinner to taste the deep, citrusy cranberry notes, or use it in the Soju Spritz (page 109) for a delightful afternoon cocktail.

In a wide-mouthed jar or glass bottle with a tight-fitting lid, combine all the ingredients. Infuse for at least 1 day, or longer for a deeper flavor.

2차

2-CHA

NO PARTY
WITHOUT AN
ANJU PARTY

You know Koreans are serious about their drinking when they dedicate an entire food group to it. This class of dishes is known as *anju*, a term that translates to "alcohol blocker" and is inclusive of all drinking food or drinking appetizers or drinking snacks—basically anything you might eat while you're drinking.

Compared to American restaurants or bars, where alcohol and cocktails lead profit margins, alcohol served in Korean drinking establishments is generally cheap. Anju is the moneymaker, and many pochas or restaurants will require an anju minimum rather than a drink minimum. Even if they don't, it's still considered rude to not order at least one dish, even if you're already on your fourth round and couldn't possibly fit another thing in your stomach.

And this is with good reason. Anju defines the Korean drinking style. It's how we can drink so much over the course of a long night, by lining our bellies with fried vegetable pancakes and fritters, laying a mortar of rice cakes or noodles between rounds, and cutting shots of soju with spoonfuls of spicy broth.

Personally, I feel like anything you eat while drinking could be classified as anju, so pursue your cravings with joy, even if that's just a bag of chips or a box of mac and cheese. But if you want to complete your Korean drinking spread, indulge in making some of my favorite anju: finger-lickin' snacks like Honey-Butter Bar Nuts (page 62), traditional Korean drinking foods like a crispy Haemul Pajeon (page 82) studded with juicy nuggets of seafood, or a rich and fiery-red Dubu Kimchi (page 73), plus my own crave-worthy anju creations like Kimchi Carbonara (page 97) and the Loaded Dwaeji Bulgogi Nachos (page 167).

LOST ON WHAT TO EAT WITH WHAT DRINK?
HERE ARE SOME GO-TO KOREAN
FOOD AND ALCOHOL PAIRINGS:

BEER	CHEONGJU	MAKGEOLLI	SOJU
Pair with dried fish and nuts (since the beer will help quench your thirst) or with Western-style delivery foods.	Pour this traditional drink alongside similarly traditional, old-school Korean dishes (known as hansik).	Drink this creamy fizzy beverage with fried foods or spicy foods; the makgeolli will temper the richness and the heat.	Cut the alcoholic bite of soju with deeply savory soups, Korean BBQ, offal, and raw fish.
Dried fish: nogari, jwipo, squid, yukpo (jerky) Roasted Half-Dried Squid with Gochujang Mayo (page 74)	**Bugak (deep-fried dried vegetables or seaweed)** Kimchi Gim Bugak (page 60)	**Fried foods: jeon, bindaetteok (mung bean pancake), buchimgae (vegetable pancake), pajeon (scallion pancake)** Haemul Pajeon (page 82)	**Stews: budae jjigae (army base stew), odeng tang (fish cake stew)** Budae Jjigae (page 169)
Nuts Honey-Butter Bar Nuts (page 62)	**Bulgogi**	**Spicy foods: dubu kimchi (tofu with stir-fried kimchi and pork), fresh kimchi** Dubu Kimchi (page 73)	**Grilled meats: samgyeopsal (pork belly), gopchang (intestines)** Samgyeopsal (page 157)
Fried chicken Chimaek Chicken (page 161)	**Jeon (pan-fried fritters)** Spam Jeon (page 85)	**Meats: pork belly** Bossam with Garlic Chive Sauce (page 93)	**Fatty foods: dakbal (chicken feet), jokbal (pig's feet)**
Pizza Korean Pizza (page 163)	**Yukhwe (beef tartare)**	**Seafood: oysters, nakji-bokkeum (stir-fried octopus)** Nakji Somyeon (page 152)	**Raw fish/shellfish: hwe (raw fish), grilled seafood** Grilled Scallops & Clams with Gochujang Butter (page 78)

When you sidle up to a bar in Korea, you might be welcomed with a few small bowls of gratis anju, like popcorn, crackers, or bar nuts. These dishes are called seobiseu 서비스, or "service," because they are literally free, a service from the establishment to welcome you in.

My version of seobiseu popcorn aims to whet your palate—something salty enough to make you thirsty, something sweet enough that you crave more, something to occupy your mouth and your hands while you drink and converse with your friends.

1. In a food processor, pulse the seaweed until reduced to tiny flakes.

2. Pop the popcorn according to its package directions.

3. Meanwhile, in a small saucepan or in the microwave, melt the butter.

4. Add the popped popcorn to a large bowl and pour the melted butter and sesame oil over the top. Sprinkle with the roasted seaweed, gochugaru, salt, and sugar and toss to combine. I like to use two bowls of the same size, invert one over the other like a clamshell, and shake them together to thoroughly mix the popcorn. Some of the seasoning might settle to the bottom, so be sure to remix or toss as you eat.

SALTY SWEET SEAWEED POPCORN
달콤 짭짤한 팝콘

Serves 3 or 4

3 (5 g) single-serving roasted seaweed packs

1 (3-ounce) pack plain microwave popcorn (no salt or butter added)

3 tablespoons unsalted butter

1½ teaspoons sesame oil

1½ teaspoons gochugaru

1½ teaspoons kosher salt

1½ teaspoons sugar

KIMCHI GIM BUGAK (FRIED SEAWEED CHIPS)

김치 김부각

Serves 2 or 3

4 full-sized sheets gim
(roasted seaweed)/nori

1 cup kimchi (pick the thinnest pieces
you can find, near the leaf edges
rather than the bottom stem ends)

12 sheets rice paper

Toasted sesame seeds

Neutral oil, such as vegetable or canola oil

Bugak is the ideal light beer snack: It's crunchy, salty, and the fresher it's made, the better. It's usually prepared by deep-frying dried vegetables or seaweed coated in chapssalpul 찹쌀풀 (glutinous rice paste), but I've deployed an easier technique here by leveraging rice paper. Thin sheets of kimchi add an extra spicy and savory layer.

1. Lay the seaweed sheets on a wire rack and spread the kimchi out on top, leaving some space in between pieces.

2. Fill a shallow bowl or plate with water. Dip a sheet of rice paper in the water, then lay it on the seaweed sheet over the kimchi. If the rice paper is round, it helps to cut it in half. Break off smaller pieces of rice paper to cover any gaps in the middle. Press in between the kimchi pieces to seal together the seaweed sheets and rice paper. Sprinkle with sesame seeds. Continue with the remaining sheets of seaweed and rice paper.

3. Leave uncovered on the counter overnight to thoroughly dry out. Once the sheets are dried and crispy, use scissors to cut the seaweed sheets around the kimchi pieces.

4. Line a sheet pan with paper towels and set near the stove. Pour 1 inch oil into a pot or pan and heat over medium-high heat until it reaches 300°F.

5. Slide the seaweed pieces in, rice paper–side up, and fry quickly, flipping once, 10 to 15 seconds on each side. Remove and drain on the paper towels.

VARIATION: USE KIMCHI SEASONING!

Seoul Sisters, a sisters-owned brand based in Korea, makes a delicious kimchi seasoning for an even easier version of this kimchi bugak. Assemble everything as directed, but skip the kimchi. After you lay down the rice paper sheets, sprinkle with 2 tablespoons kimchi seasoning, then the sesame seeds. Dry for 1 to 2 hours, then fry.

HONEY-BUTTER BAR NUTS
허니버터 견과류

2 heaping cups unsalted roasted mixed nuts (I like a combo of almonds, Brazil nuts, cashews, hazelnuts, pecans, and peanuts)

2 tablespoons unsalted butter

2 tablespoons honey

½ teaspoon gochugaru

1 teaspoon kosher salt

1 tablespoon sugar

My dad has always been a sucker for anything nuts, particularly the honey-roasted peanuts that were popular in the '90s. These nuts are inspired by the honey-butter chips and honey-butter almonds that have recently been popular in Korea. The gochugaru adds a subtle warmth, so feel free to increase or omit as you wish, or toss in more sugar if you've got more of a sweet tooth.

1. In a dry sauté pan, toast the nuts over medium heat until fragrant and slightly golden, about 3 minutes. (Even when nuts are sold roasted, they can often use additional toasting.)

2. Meanwhile, in a small saucepan over medium heat or in the microwave, melt together the butter and honey until combined. Remove from the heat and stir in the gochugaru and salt.

3. Pour the honey-butter mixture over the nuts, stirring to coat. Toast for an additional 2 minutes, stirring occasionally.

4. Remove the pan from the heat and let cool, stirring every few minutes or so.

5. Once the nuts are cool enough to handle but still warm, transfer to a heatproof bowl. Sprinkle the sugar over the nuts and toss to coat. Serve warm or at room temperature.

I'm secretly thrilled whenever I sit down at a Korean bar or pocha (street food and drink stall), and the only anju I'm given is a tiny bowl of little fishies. It's like my ideal fidget game: using my chopsticks to individually pick up each one to eat while I nod enthusiastically but only half listen to my fellow tablemates. The salty-sweetness is hopelessly addicting, and takes me back to meals of my childhood, when I would drop a few myeolchi-bokkeum (stir-fried anchovies) into the dregs of my leftover rice bowl and top it with a splash of tea.

Here, the rice oligodang (a malted rice syrup) gives a bit of glaze to the fishies, while the sugar lends just a tinge of added sweetness but not too much. Look for medium-sized fishies at the market, or feel free to sub with your preference: The very large 2-inch ones reserved for broth won't work as well in this dish, but anything smaller is fine.

STIR-FRIED FISHIES WITH PEANUTS
멸치 땅콩 볶음

1 tablespoon canola oil

2 heaping cups medium-sized dried anchovies (about 1 inch in length)

½ cup unsalted roasted peanuts

1½ teaspoons soy sauce

1 tablespoon rice oligodang (malted rice syrup)

1½ teaspoons sugar

1½ teaspoons toasted sesame seeds

1. In a wok or skillet, heat the oil over medium-low heat. Add the fishies and stir-fry for about 2 minutes, using chopsticks, a spoon, or a spatula to stir and agitate. Add the peanuts and cook for another 2 minutes. Most of the fishies will turn a little golden brown and smell toasty.

2. Add the soy sauce, oligodang, sugar, and sesame seeds. Stir to combine for another minute, then remove from the heat. These can be served warm or at room temperature.

SOY SAUCE JALAPEÑO PICKLES

간장 할라피뇨 피클

Makes 1 cup

1 cucumber, sliced into rounds ⅛ inch thick

1 large jalapeño, sliced into rounds ⅛ inch thick

4 garlic cloves, peeled but whole

2- to 3-inch strip lemon zest, removed with a vegetable peeler

⅓ cup soy sauce

¼ cup rice vinegar or distilled white vinegar

2 tablespoons sugar

I've been making these salty-sweet pickles with a mix of crunchy cucumbers and spicy jalapeños ever since my first Yooeating pop-up, and they've adorned everything from Korean Cubanos to tostones nachos to our anju menu at Orion Bar. They're the perfect fridge stash for any late-night drinking sesh. The salty pickled cucumbers are lip-smackingly good with a cold beer, while a slice of the hot pepper can be tamed with a swig of creamy makgeolli. Put them together for an eye-opening punch that's assertive enough to stand up to a shot of soju.

1. In a heatproof, lidded container (like a 24-ounce mason jar), combine the cucumbers, jalapeño, garlic cloves, and lemon zest.

2. In a small saucepan, combine the soy sauce, ⅓ cup water, the vinegar, and sugar. Bring to just below a boil over medium heat (small bubbles will start to form), stirring to dissolve the sugar.

3. Remove from the heat and pour over the cucumber and jalapeño, pressing them down until they are fully submerged. Allow to cool, cover the container with the lid, and let steep overnight in the refrigerator.

4. Eat as an anju with soju, a banchan with rice, or on the Loaded Dwaeji Bulgogi Nachos (page 167). These pickles will keep for several weeks.

CORN CHEESE
QUESO →

← SHRIMP CHIPS &
MYEONGNANJEOT
DIP

SHRIMP CHIPS & MYEONGNANJEOT DIP

새우깡과 명란젓 딥

Makes ⅔ cup or serves 4

2 sacs salted and fermented pollack roe (sold as myeongnanjeot or mentaiko); if there are 2 lobes attached to each other, each counts as 1 sac

½ cup mascarpone or sour cream

½ teaspoon grated lemon zest

1 teaspoon fresh lemon juice

2 scallions, minced

Tobiko or salmon roe (optional)

1 (2.6-ounce) bag shrimp crackers, for serving

The iconic red bag of Nongshim's Shrimp Crackers 새우깡 (Saeukkang), introduced in 1971, has been a staple snack for me since my childhood. Like many snack foods of that era, the cracker is modeled after a similar Japanese shrimp chip by Calbee called Kappa Ebisen but became such a huge hit in its own right that it's now found worldwide. If you've never had it (where ya been?), Shrimp Crackers are crispy little pinky finger-sized crackers that smell and taste faintly of sweet shrimp, which become only more addictive the more you munch on them.

On the other hand, salted and fermented pollack roe (myeongnanjeot 명란젓) is incredibly popular in Japan (where it is known as mentaiko), though it's actually of Korean origin. It's usually served as a banchan—I could eat a whole bowl of rice with it—and its creamy brininess gives a subtly indulgent and irresistible flavor to this dip.

1. Pierce the pollack roe sacs with a sharp paring knife and use the back of the knife, clean hands, or gloved hands to scrape or squeeze out the inside roe. Transfer to a small serving bowl. Stir in the mascarpone, lemon zest, lemon juice, and scallions. Mix thoroughly to combine.

2. Dollop a spoonful of tobiko or salmon roe on top, if desired. Serve with shrimp crackers.

I once had a roommate who hailed from Texas, and she was always on the hunt for the perfect queso. We eventually found it at the backyard taco truck at Union Pool, a bar in Williamsburg, where we'd get our Saturday drink on early and wait patiently with Shiner Bocks until they opened at 1 p.m. so that we could chow down on some tacos and a cup of their queso. It hit all the right notes—silky, creamy, and salty enough that we kept having to order more beers (and an extra side of chips).

My Korean American queso incorporates corn cheese, a dish that's served at Korean BBQ restaurants (often made directly in the well of the grill pan, where it melts as you cook your meat). I like to use white American cheese for my queso (you can use yellow as well); it melts to the perfect consistency and holds up to rounds of reheating over the course of the night. Make sure to load up on lots of chips—why are there never enough?

CORN CHEESE QUESO
콘치즈 퀘소

Makes 2 cups

2 tablespoons unsalted butter

2 tablespoons all-purpose flour

⅓ cup whole milk

2 garlic cloves, minced

1½ cups corn kernels

3 scallions, chopped

¾ teaspoon kosher salt

¼ teaspoon freshly ground black pepper

1¼ teaspoons gochugaru

½ pound white American cheese, sliced (about 1½ cups if shredded)

½ cup shredded mozzarella cheese

Lots of thick tortilla chips, for serving

1. In a small pot or saucepan, melt the butter over low heat. Add the flour and stir together using a wooden spoon or silicone spatula until some of the moisture evaporates and the mixture smells nutty, 5 to 8 minutes. This roux should look golden but not too browned.

2. Add a bit of the milk and stir it in to temper any splashes or bubbles, then add the rest slowly. Stir in the garlic, corn, scallions, salt, pepper, and 1 teaspoon of the gochugaru and cook for 5 minutes.

3. Reduce the heat to low and stir in the cheeses. Once the cheese melts, remove the queso from the heat. Transfer to a heatproof serving bowl or warming dish and garnish with the remaining ¼ teaspoon gochugaru. Serve with tortilla chips.

When I polled my group of friends (and a bunch of strangers on the Internet), this dish was one of the top-requested anju. It's a simply named dish that marries soft bouncy dubu (tofu) with a strong punchy kimchi through the power of pork, making a harmonious match for both soju and makgeolli.

The other advantage to this dish is that it's an easy fridge-cleanout dish. Use up that old or super-sour kimchi languishing in the back of your fridge, and you can also sub in pork shoulder for the pork belly in a pinch. I use a Paik Jong-won (South Korea's Gordon Ramsay–meets–Julia Child) tip here by adding a bit of brown sugar to the pork belly, which makes a kind of meaty caramel reminiscent of good burnt-end BBQ.

1. Place the tofu in a small pot, add 1 teaspoon of the salt and water to cover. Heat over high heat until the water is just about to boil, then turn off the burner. Set aside.

2. In a wok or skillet, heat the oil over medium-high heat. Spread the pork belly slices in the pan and cook for about 3 minutes on one side, then sprinkle half the brown sugar over them, flipping once the sugar has melted. Add the rest of the brown sugar and cook until both sides of the pork belly are golden brown, flipping as needed, another 3 to 4 minutes.

3. Remove and reserve the pork belly and spread the kimchi over the pork fat that's pooled in the skillet. Cook for about 3 minutes, flipping once. Add the remaining ¼ teaspoon salt, the garlic, and the gochugaru and stir in the pork belly. Cook until the kimchi and pork belly are a deep red with some charred bits on the edges, another 4 to 5 minutes. Remove from the heat and stir in the sesame oil.

4. Carefully drain the tofu. Cut in half lengthwise and then into ½-inch-thick slices. Fan out the slices on one half of a plate and drizzle a little sesame oil over the top. Plate the pork and kimchi alongside. Sprinkle sesame seeds over everything and serve immediately.

DUBU KIMCHI (TOFU WITH STIR-FRIED KIMCHI AND PORK)
두부김치

Serves 2 as a main, 4 as an appetizer

1 (14- to 16-ounce) box medium-firm tofu

1¼ teaspoons kosher salt

1 tablespoon canola oil

1 cup thinly sliced pork belly (1-inch squares about ¼ inch thick)

1 tablespoon brown sugar

1 cup kimchi, cut into about 1-inch squares

1½ teaspoons minced garlic

2¼ teaspoons gochugaru

1 teaspoon sesame oil, plus more for garnish

1 teaspoon toasted sesame seeds

ROASTED HALF-DRIED SQUID WITH GOCHUJANG MAYO

고추장 마요네즈 반건조 오징어

Serves 2 or 3

1 whole half-dried squid

2 tablespoons Kewpie mayo

1 tablespoon gochujang

Juice of ½ lemon

A few years back, I traveled to Jeju Island with my parents, where we drove around visiting landmarks and eating good food. One of the most memorable moments was getting out of the car at the beach in Seopjikoji and immediately getting hit with the aroma of squid roasting on coals; mixed with the brisk ocean air, it was intoxicating. My mom and I ran over to the stall flanked with rows of squid drying on clotheslines and squealed like two schoolgirls as we picked out the best-looking squid. We grabbed a cold beer from the tiny mart next door and settled into some plastic chairs along the seawall, passing the can back and forth as we happily chewed on our perfectly grilled squid tentacles.

To re-create the experience at home (sans the Jeju beach air), you can seek out dried squid in the frozen seafood section of Korean grocery stores. I prefer half-dried squid for a better texture, but if you have fully dried squid, you can boil it for 30 to 40 seconds or soak it in soju for 10 minutes to rehydrate a bit. While you'll see dried squid commonly served with Cheongyang chili pepper mayo seasoned with soy sauce, I like the bright spiciness of this gochujang mayo.

1. Use kitchen scissors to make 1-inch incisions into each side of the squid body.

2. In a small bowl, stir together the mayo, gochujang, and lemon juice. Set aside.

3. Grill the squid over a flame on low heat (I just use the gas burner of my stove or a portable gas stove), turning often, for about 10 minutes until the squid is heated through and charred along the edges. (Alternatively, you can grill in a pan with a tablespoon of butter over medium heat, for 8 to 10 minutes.)

4. Cut the squid body into small strips and tear the tentacles apart. Serve with the gochujang mayo.

DRIED FISH ANJU 마른안주*

Since Korea is small and surrounded by ocean, fish—particularly dried fish—has long been a huge part of our culture. Of course, drying preserves the fish, but it also concentrates those delicious glutamates. Below are some common dried fish anju, which you can find packaged at Asian grocery stores in plastic clamshells or vacuum-sealed bags, sometimes seasoned. Feel free to stir-fry with your own seasonings.

Jwipo 쥐포

A pressed, usually round fish jerky made with filefish that's kind of sweet and candied

Pollack, or myeongtae 명태

A fish that's so popular and prevalent in Korea that there are over thirty alternative names for it, depending on its age and frozen or dried form, including:

Myeongtae-po 명태포: dried pollack

Nogari 노가리: dried young pollack

Hwangtae-chae 황태채: dried and sliced yellow pollack

Ojingeo-po 오징어포 or Ojingeo-chae 오징어채

Dried or dried-and-sliced squid

Gwamegi 과메기

A half-dried version of Pacific saury, an oily fish known as kkongchi 꽁치, that's popular in Pohang and served with scallions and roasted seaweed as a ssam (wrap)

* The term 마른안주 usually denotes any dry anju, including nuts, puffed rice snacks, dried fruit, or even candy.

GRILLED SCALLOPS & CLAMS WITH GOCHUJANG BUTTER

고추장 버터 가리비 조개구이

Serves 2

SHELLFISH
12 clams, scrubbed
2 tablespoons kosher salt
¼ cup soju (optional)
4 live scallops in the shell

SAUCE
1 tablespoon gochujang
1 tablespoon minced garlic
1 tablespoon unsalted butter, melted
¼ cup shredded mozzarella cheese
1 Cheongyang chili pepper or serrano chili pepper, seeded and minced
1 scallion, minced

Whenever I'm traveling along the southern coast of Korea and get a whiff of salt water, I crave grilled shellfish and soju. It was one of the things I most looked forward to when visiting my friend Jamesy in his hometown of Pohang, where the beaches are lined with pochas (food and drink stalls) that specifically serve seafood. Huge blue tanks filled with fresh sea squirt, long-limbed crabs, and nubby-legged octopus flanked the entrance to Eun-jung's Grilled Clams 은정이네 조개구이, a spot I picked mainly because it had the same name as my sister.

What followed was everything I had dreamed of: fresh Dokdo dohwa shrimp with a side plate of their heads fried crispy, a tower of seafood on the half shell adorned with shredded cheese and flecks of spicy pepper, an aluminum tin of spicy seafood ddukbokki, a metal basin of abalone ramyun. We cracked open cold beers and poured each other soju while the ajumma took great care to show us how to grill the seafood and when to add a little squirt of clam juice (right when you place them on the grill, her secret trick).

It's not easy to re-create this experience at home, especially as someone who lives in a fourth-floor walk-up in Brooklyn, miles from the sea and without access to a charcoal grill. But this seafood platter, gilded with schmears of gochujang butter made bubbly under the broiler, can transport you right to the rocky seashores of southern Korea.

1. Prep the clams Submerge the clams in a bowl with 2 cups water mixed with the 2 tablespoons salt. Store in the fridge for at least 1 hour to purge any grit and sand, then rinse.

2. Shuck the clams on the half shell or steam them open by adding them to a shallow pan with 1 cup water and ¼ cup soju over medium heat, removing each clam as it opens up, about 5 minutes. (Discard any clams that do not open.) Reserve ¼ cup of the steaming liquid if available.

3. Prep the scallops Use a thin knife to open the scallop shells from mouth to hinge. Remove the flat top shell. Remove and discard the dark digestive gland from the side of the first scallop and the brown frilled gill in between the scallop and the outer mantle ring. Preserve the scallop, its pink roe sac if present, and the mantle. Loosen the scallop from the shell, rinse in cold water, then slice horizontally into 3 or 4 thin →

rounds. Return the pieces to the shell. Repeat with the rest of the scallops.

4. Make the sauce In a small bowl, mix together the gochujang, garlic, and melted butter until combined. Dollop a bit of sauce on each clam and spoon a larger amount on the scallops. Sprinkle the cheese, chili pepper, and scallion on top.

5. To cook on a grill Set a charcoal or gas grill to high heat. Set the scallops and clams on a grill topper. Grill, covered, until the seafood is cooked through, the edges are slightly charred, and the cheese is melted, about 8–10 minutes. Add a bit of the clam steaming liquid or more butter if any shells get too dry as they cook.

  Or, to cook under the broiler Arrange scallops and clams in a single layer on a sheet pan. Broil until the seafood is cooked through, the edges are slightly charred, and the cheese is melted, checking every few minutes and adding a bit of the clam steaming liquid or more butter if any shells get too dry.

HAEMUL PAJEON (SEAFOOD PANCAKE)
해물파전

Makes 2 pancakes (serves 3 or 4)

PAJEON
3 to 5 scallions (depending on thickness)

¾ cup all-purpose flour

1 tablespoon cornstarch

1 tablespoon soy sauce

Salt and freshly ground black pepper

¼ cup vegetable oil

1 cup frozen seafood mix (no need to thaw)

DIPPING SAUCE
2 tablespoons soy sauce

1 tablespoon rice vinegar

½ teaspoon toasted sesame seeds

½ teaspoon gochugaru

When it rains, the soft pitter-patter of drops hitting the ground conjures up cravings for makgeolli and this classic anju, a crispy-edged pancake built on beams of pungent scallions and studded with plump bits of seafood. The rain is reminiscent of the sound of the sizzle of the pancake frying in the oil (romantic, right?).

1. Make the pajeon Cut the scallions in half lengthwise, and again if quite thick. Cut in half crosswise so they will fit a nonstick medium pan.

2. In a medium bowl, mix together flour, cornstarch, 1 cup water, and soy sauce into a runny batter. Add a pinch of salt and pepper.

3. Line a plate with paper towels and keep it near the stove. In a shallow nonstick medium pan, heat half of the oil over medium-high heat until shimmering, 1 to 2 minutes. Add ⅔ cup of the batter and swirl the pan to spread into a thin layer. Spread half the scallions evenly in a row and stud the pancake with half the seafood, pressing them down into the batter with a spatula. Drizzle on another ⅓ cup of the batter. Cook until golden brown underneath, about 5 minutes total. Then flip and cook until the edges are crispy and the batter is cooked through, another 3 to 5 minutes. Transfer to the paper towel. Add the remaining oil and repeat with the remaining batter, scallions, and seafood to make a second pancake.

4. Make the dipping sauce In a small bowl, mix together the soy sauce, vinegar, 1 tablespoon water, the sesame seeds, and gochugaru.

5. Cut the pancakes into roughly 2-inch squares (you can use kitchen scissors) and serve with the dipping sauce.

Ah, Spam, that gorgeous blue tin with the pink canned meat. The can alone evokes a ton of nostalgia: the crack of the pull-tab top, the thwack against the cutting board as the ham shimmies out with a satisfying plop. It's beloved in South Korea: simmered in Budae Jjigae (page 169), gift-wrapped for major holidays, and served as an anju at various pochas (street food and drinking stalls), sliced and pan-fried with a squeeze of ketchup.

I first fell in love with Spam as a kid, when my mom would coat a few slices in flour and egg and fry it up as an after-school snack. This simple batter technique can be applied to any meat or vegetable to make jeon, a pan-fried fritter that's a common anju. While I used to eat my mom's Spam jeon over rice with a few shakes of furikake, these days I'll have it with a blue bottle of Jinro Is Back.

SPAM JEON
스팸전

Serves 4

1 can Spam
¼ cup all-purpose flour
2 eggs
2 tablespoons canola oil

1. Cut the Spam lengthwise into 8 slices and then cut each slice in half crosswise to create 16 squares total. (If you plan to use these in the Spam Jeon Musubi, page 86, you can leave the Spam cut into 8 slices. Save the Spam can to use as a musubi mold if you don't have one for the Spam Jeon Musubi recipe.)

2. Set up a dredging station in two shallow bowls or plates: Spread the flour in one bowl and in the other whisk the eggs.

3. In a large skillet, heat 1 tablespoon of the oil over medium-high heat. Coat half of the Spam slices in the flour and shake off any excess. Dip in the egg, making sure to cover both sides, and let the excess drip off. Fry the Spam until golden brown on both sides, 2 to 3 minutes per side. Repeat with the remaining Spam.

To serve: Plate with a swirl of ketchup or mayo as an anju, or eat with rice as a banchan. Or, better yet, reserve to make Spam Jeon Musubi (on the next page).

SPAM JEON MUSUBI
스팸전 무스비

Makes 8

8 Spam Jeon (page 85)

1 to 2 teaspoons vegetable oil

2 tablespoons unsalted butter

½ cup kimchi, roughly chopped
into small pieces

MUSUBI

4 (8 × 8-inch) sheets gim
(roasted seaweed)/nori,
cut into 3-inch-wide strips

4 cups cooked white rice

4 teaspoons furikake

4 slices American cheese, halved

You'll commonly see Spam musubi at Asian convenience stores or as a must-have item on Hawaiian menus—a slab of fried Spam atop a bed of pressed white rice tucked in a seaweed sheet sleeve. I've made this amped-up version for Yooeating pop-ups and summer picnics, and it now has a permanent home on the anju menu at Orion Bar. With the addition of layers of caramelized kimchi, furikake, and just-slightly-melted cheese, it's like a portable, handheld version of a very addictive kimchi fried rice.

If you don't have a musubi mold (which you can find easily at Asian grocery stores or online), you can just use the Spam can itself!

1. Reheat the Spam jeon in a nonstick skillet over low heat with a little bit of oil, 1 to 2 minutes on each side. Remove from the heat but leave the jeon in the pan to keep warm.

2. In a small nonstick pan, melt 1 tablespoon of butter and add the kimchi. Cook, stirring once every minute or two. The moisture of the kimchi will cook off and the kimchi will begin to caramelize. After about 5 minutes, or when the pan feels dry, add the second tablespoon of butter. Cook until the kimchi has reduced in volume by half, is nicely browned and fragrant, and the edges are slightly charred, another 3 to 5 minutes. Remove from the heat and set aside.

3. Build the musubi Lay a strip of seaweed on your work surface or cutting board with one short end facing you. Position the musubi mold (or Spam can!) perpendicular to the strip and near its top edge, leaving a bit of space on the seaweed above. Layer ½ cup of rice into the mold, using the mold press or a spoon to spread and press evenly. Sprinkle ½ teaspoon furikake over the rice. Add a half slice of cheese, then about 1 teaspoon kimchi, and finally top with one Spam jeon. Press together with the mold press and remove from the mold (or if you're using the Spam can, invert gently onto the seaweed and shimmy the stack out). Carefully roll the musubi along the seaweed strip to wrap evenly, making sure the gim is folded around each edge. Dab a bit of water along the edge of the gim to seal it or use a kernel or two of rice. Repeat with the remaining ingredients to make 7 more musubi.

4. Cut the musubi in half to serve if desired.

For me, the bright red spicy ddukbokki (stir-fried Korean rice cakes) is like the Korean Kraft Mac and Cheese: comforting and nostalgic for kids and adults alike. My mom would take us as young kids to the bunsik (street food) stall she used to frequent in college for a ddukbokki fix, washing the rice cakes in a bit of water for us to remove some of the spiciness. As we grew up, we learned to seek out that sticky, sweet spice as a family, and later as tipsy adults when we were out late drinking.

Sure, you can steep some kelp and anchovies to make your own broth base at home, but to be honest, late-night drinking food is not the place to spend time getting fancy. Dasida, an umami-rich beef- or anchovy-flavored soup stock powder, is the secret here to re-create that simple and comforting childhood taste. It's as easy as tearing open a blue box of mac and cheese.

1. Rinse the rice cakes in water and set aside. Cut the fish cake sheets into quarters, then cut them in half again diagonally to create triangular pieces.

2. In a medium wok or skillet, combine the oil, gochujang, sugar, and soy sauce and stir-fry over medium heat for 2 to 3 minutes to loosen the gochujang and combine the sauce.

3. Add 1¾ cups water and increase the heat to high. Add the gochugaru, rice syrup, Dasida, garlic, and rice cakes and bring to a boil. Reduce the heat to medium-low (so the sauce is lightly bubbling) and simmer until the sauce thickens and the rice cakes are soft and plump, 8 to 10 minutes, stirring regularly so the rice cakes cook evenly and don't stick.

4. Add in the 2-inch pieces of scallion and the fish cakes and cook for another 2 to 3 minutes.

5. Remove the pan from the heat and top with the mozzarella. Garnish with some sesame seeds and the thinly sliced scallion. Squeeze the lemon wedge over the top and serve immediately.

Note: I use jocheong 조청, a type of Korean rice syrup, in this recipe, but you can also use another syrup, such as oligodang 올리고당 (malted rice syrup) or mulyeot 물엿 (corn syrup).

CHEESY SPICY DDUKBOKKI
매콤한 치즈떡볶이

Serves 2 to 4

18 ounces (500 g) Korean rice cakes (dduk)

2 sheets thin frozen pre-fried fish cakes (eomuk)

1 tablespoon vegetable or canola oil

2 tablespoons gochujang

1 tablespoon sugar

1 tablespoon soy sauce

2 tablespoons fine gochugaru

2 tablespoons rice syrup (jocheong; see Note)

1½ teaspoons Dasida soup stock powder (beef or anchovy flavor)

1 tablespoon minced garlic

2 to 3 scallions, halved lengthwise and cut crosswise into 2-inch pieces, plus 1 scallion, thinly sliced

4 ounces mozzarella cheese, shredded

Toasted sesame seeds, for garnish

1 lemon wedge

WHITE DDUKBOKKI
하얀 떡볶이

Serves 2

16 ounces Korean rice cakes
(about 2½ cups)

2 sheets thin frozen pre-fried
fish cakes (eomuk)

¼ medium yellow or white onion,
cut into slices ⅓ inch thick

4 shiitake mushrooms, stems
removed, caps sliced ⅓ inch thick

3 scallions, cut on the diagonal
into 1-inch lengths

½ to 1 Cheongyang chili pepper
(or serrano chili pepper)

1 tablespoon minced garlic

1 tablespoon sugar

1½ teaspoons Dasida soup stock
powder (beef or anchovy flavor)

1 teaspoon soy sauce

½ teaspoon kosher salt

½ teaspoon freshly ground black pepper

Toasted sesame seeds, for garnish

You know a place is good when three different people take you there on different occasions, independently of each other. That was the case with Neunejib 느네집 in Itaewon, a little alley spot that specializes in a dish they call "white ddukbokki." Their stir-fried rice cakes don't have any of the classic gochujang or gochugaru found in the saucy red version, but it still packs a spicy punch from the Cheongyang chili peppers as well as a rich savoriness from the mushrooms and beef-flavored broth. It's kind of similar to the soy sauce–based gungjung ddukbokki, also known as royal court ddukbokki, a dish that dates back to Korea's Joseon dynasty and that my mom used to make every year for my dad's birthday party, the one time we'd ever see him drink.

1. Rinse the rice cakes in cold water. Drain and set aside. Cut the fish cake sheets into quarters, then cut them in half again diagonally to create triangular pieces.

2. In a wide shallow pot or pan, combine the onion, mushrooms, scallions, rice cakes, fish cakes, Cheongyang chili pepper, garlic, and 1½ cups water. Bring to a boil over high heat. Reduce to medium heat. Stir in the sugar, Dasida, soy sauce, salt, and black pepper and cook, stirring frequently, until the rice cakes are soft and plump but not mushy, about 5 minutes.

3. Remove from the heat, sprinkle with sesame seeds, and serve immediately.

I always struggle with whether to call this classic Korean pork belly dish "bossam" or "suyuk." A bit of that is attributable to regional differences (I have never heard my family refer to it as anything other than bossam*), but technically,* suyuk *refers to this technique of cooking pork belly, while* bossam *refers to the pork belly wrap as a whole.*

Either way, this bouncy, succulent pork belly makes for an excellent anju, classic for pairing with both creamy makgeolli to soothe the intense heat of the raw garlic chive sauce, and with crisp soju to cut through the fattiness of the pork.

1. Slice the pork belly into 3-inch-by-3-inch wide logs if it isn't already divided into portions.

2. In a heavy-bottomed pot or Dutch oven, heat the oil over medium-high heat until shimmering. Add the pork belly and sear until golden brown on all sides, 2 to 3 minutes per side. Remove and set aside.

3. Add the onion, garlic, and leek to the pan and sear until charred and browned on all sides, for 2 to 3 minutes per side.

4. Add 8 cups water, the doenjang, soy sauce, peppercorns, instant coffee, apple, and soju. Bring to a boil. Reduce to a simmer and return the pork belly to the pot. Cook, uncovered, until tender, about 45 minutes. The juices will run clear when pierced with a knife, and both the fat and meat should be jiggly.

5. Turn off the heat and store the pork belly in the broth until you're ready to serve it, to keep it moist. When ready, remove the pork and slice thinly.

6. Serve the pork belly slices with the garlic chive sauce, kimchi, and red leaf lettuce for wrapping.

Garlic Chive Sauce • Makes ½ cup

In a blender or food processor, combine the garlic chives, salt, gochugaru, sugar, sesame oil, vinegar, soy sauce, sesame seeds, garlic, and pepper to taste and blend. With the machine running, stream in the olive oil until it's well incorporated and the sauce is smooth.

BOSSAM (BOILED PORK WRAPS) WITH GARLIC CHIVE SAUCE
마늘 부추 소스 보쌈

Serves 2 to 4

PORK BELLY

2 pounds pork belly (whole or logs, ideally with the skin on)

1 tablespoon canola oil

1 onion, quartered

10 whole garlic cloves, peeled

1 leek, halved lengthwise and cut crosswise into 2-inch pieces

¼ cup doenjang

2 tablespoons soy sauce

1½ teaspoons black peppercorns

1 tablespoon instant coffee

½ apple (I like Fuji), cored

¼ cup soju

FOR SERVING

Garlic Chive Sauce (recipe follows)

Kimchi

Red leaf lettuce, for wrapping

Garlic Chive Sauce

1 cup (4 ounces) roughly chopped garlic chives

¾ teaspoon kosher salt

2 teaspoons gochugaru

¼ teaspoon sugar

1½ teaspoons sesame oil

1½ teaspoons rice vinegar

1½ teaspoons soy sauce

1½ teaspoons toasted sesame seeds

6 garlic cloves, peeled but whole

Freshly ground black pepper

¼ cup olive oil

JOGAETANG (SPICY CLAM SOUP)

조개탕

Serves 2 to 4

1½ pounds littleneck clams
(16 to 17 clams)

2 tablespoons plus
½ teaspoon kosher salt

1 (6-inch) piece dasima
(dried kelp)/kombu

6 dried anchovies for broth
(2 to 3 inches long), guts removed

½ to 1 Cheongyang chili pepper or
serrano chili pepper, thinly sliced

1 garlic clove, minced

½ teaspoon fish sauce

½ teaspoon mirin

Freshly ground black pepper

¼ wedge lemon

Jogaetang is my ultimate anju, the very specific thing I crave when a shot of soju is placed in front of me. The brininess cuts through the soju, and the spicy pepper infused into the soup hits you in the back of your throat, clearing everything else out and making you go "Ahhh" after each spoonful. It's the epitome of the Korean concept of siwonhada, *something that is hot and refreshing and nourishing.*

The basis of this dish is a good myeolchi yuksu (anchovy broth), and then the clams do the rest of the work.

1. Soak the clams in a large bowl with 2 cups water and 2 tablespoons of the kosher salt mixed in; it should taste salty like the sea (if not, add more salt). Store in the fridge for at least 1 hour. Drain and rinse when ready to use.

2. In a saucepan, combine the dasima, anchovies, and 4 cups water. Bring to just below a boil over medium heat. Remove the dasima (save for another use, like "Cold Brew" Leftover Dasima Broth, see page 121) and reduce the heat to medium-low. Simmer for 10 to 15 minutes (little bubbles should be breaking the surface) until the broth is fragrant and the anchovies start to disintegrate. Discard the anchovies.

3. Increase the heat to medium. Add the remaining ½ teaspoon kosher salt, the chili pepper, and garlic and cook for 1 to 2 minutes until fragrant. Add the clams and cook, stirring occasionally, until they open, about 10 minutes. Remove from the heat and discard any clams that are broken or clearly not opening. Stir in the fish sauce, mirin, and black pepper.

4. Serve with the lemon wedge. Chase shots of soju with spoonfuls of the soup.

This is the recipe that most represents the type of food I like to eat when I'm drinking. It's savory and fatty, it's got noodles, and it comes together in 10 to 15 minutes. Best of all, I almost always have the ingredients for it on hand or can easily sub as needed. I can make it when I'm drunk, I can make it when I'm hungover. It's a recipe that I've developed over and over, for my YouTube channel, for food media outlets, for the bar, and even for other people's cookbooks.

Here, in my own cookbook, is the be-all and end-all version. I like to use Sun ramen noodles, used in many Japanese ramen restaurants, because their alkaline noodles make for the perfect chewy texture in the sauce, and they're hard to overcook (important when making this dish drunk).

This dish is based on a classic Roman-style carbonara, not the carbonara that's most commonly found in Korean fusion cooking, with pours of heavy cream and dots of green peas. For that reason, the sauce is built in the serving dish itself. Enlist your eating partner to stir their own noodles alongside you for maximum efficiency and eating ease.

1. Bring a pot of water to a boil. (I don't salt the water because Sun ramen noodles are alkaline and the dish itself is already pretty salty, but if you are using other noodles, definitely salt your water once it comes to a boil.)

2. In a skillet, cook the bacon over medium heat until the fat is rendered but the pieces are not yet too crispy, stirring occasionally, about 4 minutes. Transfer the bacon and its fat to another bowl, leaving about 1 tablespoon of fat in the pan.

3. Add the kimchi to the pan and cook over medium heat until the moisture is cooked out and the edges become lightly charred and caramelized, stirring occasionally, 5 to 8 minutes. Transfer the kimchi to the bowl with the bacon.

4. While the kimchi is cooking, prepare two pasta bowls. Into each bowl, crack 1 egg and add ¼ cup Parmesan, and ¼ teaspoon black pepper. Whisk with a fork to combine thoroughly.

5. Boil the noodles according to package directions (2 minutes for Sun ramen). Reserving ¼ cup of the pasta water, drain the noodles. →

KIMCHI CARBONARA
김치 카르보나라

Serves 2

2 portions (10 ounces) fresh ramen noodles (Sun ramen, if you can find them) or 8 ounces spaghetti

3 slices thick-cut bacon, diced

½ cup (packed and heaping) kimchi, chopped

2 large eggs

½ cup freshly grated Parmesan cheese, plus more for serving

½ teaspoon freshly ground black pepper, plus more for serving

2 scallions, chopped

6. Divide the noodles and the bacon-kimchi mixture equally between the two bowls. Work quickly to toss the noodles in the sauce, turning and coating each strand, until the sauce thickens. You can add a splash of pasta water if the noodles begin to feel dry.

7. Top the noodles with more Parmesan, black pepper to taste, and the chopped scallions (it will seem like a lot at first glance but will balance out once mixed in).

It might seem strange to include a recipe for instant ramyun in this book. After all, the package itself comes with directions. But I think the beauty of Korean ramyun is that, though at its core it's just a simple bag of deep-fried noodles, with a couple packets of sauce or dried veggie flakes, it can be the foundation for so much more.

Truffle jjapagetti and its cousin jjapaguri (an amalgamation of jjapagetti and the Neoguri seafood ramyuns from Nongshim famously featured in the Oscar-winning movie *Parasite*) are increasingly being served in modern pochas and Korean restaurants. Though these versions often feature cubes of steak as a nod to the movie, I like mine with thinly sliced pieces of beef brisket called chadolbagi, which cook up quick and wrap neatly around each chopstickful of black bean–sauced noodles.

1. In a medium pot, combine the sesame oil, brisket slices, and the salt and pepper. Flip or stir quickly over medium heat (the sesame oil can burn easily and the brisket will cook super fast). Once cooked through and slightly crispy on the ends, about 1 minute, remove the brisket from the pot and set aside.

2. Add the scallion whites to the pot and stir-fry them in the residual oil over low heat until fragrant to make a quick scallion oil, 1 to 2 minutes.

3. Add a generous 2 cups water to the pot and increase the heat to high. Add the vegetable flake and sauce powder packets from the ramyun package. (You can discard the olive oil packets or reserve them for another use.) Once the water is boiling, add the noodles and cook for 3 to 4 minutes. The noodles may not be completely submerged immediately, so flip them as needed until they soften and sink.

4. Remove the pot from the heat. I like to serve the ramyun directly in the pot, but you can also transfer it to a serving dish. Layer the brisket on top, and then sprinkle with the truffle oil, the scallion greens, and the sesame seeds. Top with julienned cucumber and serve immediately.

Note: If you can't find Chapagetti-brand noodles, get another similar black bean sauce–flavored (jjajang-flavored) instant ramyun, such as Chacharoni or Zha Wang.

TRUFFLE JJAPAGETTI WITH BEEF BRISKET
차돌박이 트러플 짜파게티

Serves 2

¼ teaspoon sesame oil

4 ounces brisket, thinly sliced

Pinch of kosher salt and freshly ground black pepper

2 scallions, thinly sliced, whites and greens separated

2 (4.5-ounce) packs Chapagetti jjajang noodles (see Note)

1 tablespoon truffle oil

1 teaspoon toasted sesame seeds

¼ cup thinly julienned cucumber

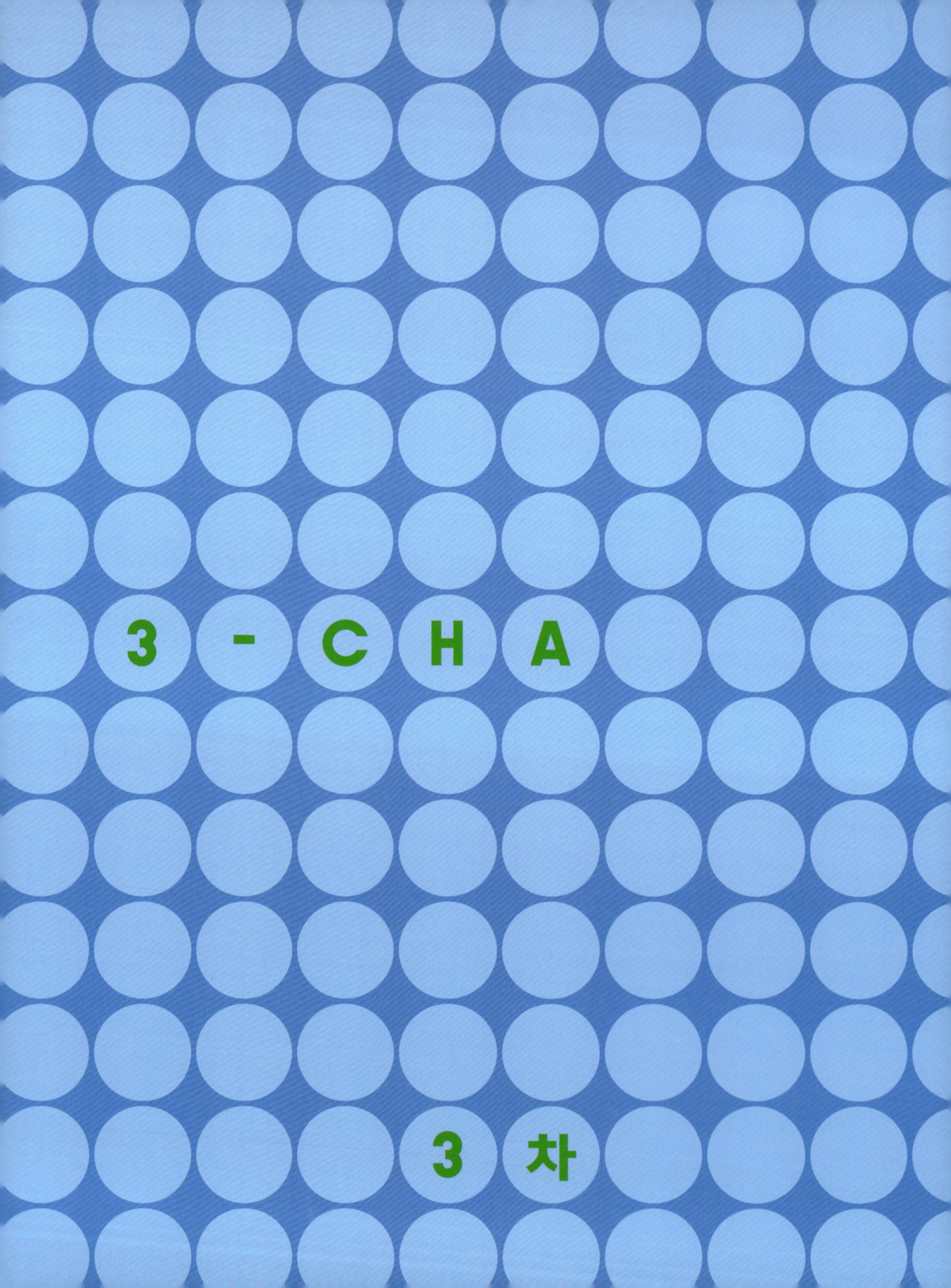

3 - CHA

3 차

KOREAN

(MEETS)

AMERICAN

COCKTAILS

RIFFIN'
ON THE
CLASSICS

When Nick and I developed the drinks menu for Orion Bar, we wanted to highlight Korean alcohol and ingredients through the lens of classic American cocktails.

There is an unspoken short list of these classic cocktails you can order at most bars, even if it's not explicitly written on their menus. As you learn to drink, you can try different versions of these cocktails to gain an understanding of what you like before you settle into something you know, a reliable order for when the bartender comes around even when you're at an unfamiliar place. For me it's a well-balanced Negroni; for others it's a sharp martini or a heady old-fashioned.

In Korea, these types of drinks are slowly gaining traction as Koreans expand their palate beyond soju into foreign spirits. But at Orion Bar we wanted to break the mold for how Korean alcohol is traditionally served and drunk. We mixed soju into salty martinis tinged with seaweed and mushroom and poured michelada mix over crisp Korean beers. Korean sool (alcohol) like maesil-ju (green plum wine) stood in for liqueurs like triple sec, and we created our own amaro-inspired herbal liqueur with dried jujubes and omija. The end result are cocktails that are appealing to anyone, whether you're sipping familiar Korean flavors in a fresh context or exploring new flavors against a tried-and-true backdrop.

Just like at a bar, start with your go-to cocktail, and then try some others. Who knows? Maybe you'll unlock a new favorite.

PERILLA YUJA SOJU & TONIC
깻잎 유자 소주 토닉

Makes 1 cocktail

Ice

1½ ounces soju

½ ounce yuja (yuzu) juice

¼ ounce Perilla Simple
Syrup (recipe follows)

Perilla leaf, for garnish

3 ounces high-quality tonic water

Perilla Simple Syrup

1 cup sugar

6 medium perilla leaves

If I could sip this year-round, I would drink nothing else, a riff on the ever-classic gin and tonic. By using soju as the base liquor, the citrusy tartness of the yuja (yuzu) fruit and the grassy anise flavors of the perilla (kkaennip) leaf really shine.

At the bar, we make our own tonic and carbonate the drink in-house to infuse the cocktail with bubbly effervescence on the spot. At home you can use high-quality tonic water for a similar effect.

In an ice-filled cocktail shaker, combine the soju, yuja juice, and perilla simple syrup and shake vigorously for 20 to 30 seconds to combine. Place a perilla leaf along the side of an ice-filled highball glass. Add tonic water and strain the cocktail over top. Stir gently and serve.

Perilla Simple Syrup • **Makes 1½ cups**

In a small saucepan, combine 1 cup water and the sugar and stir over medium heat until the sugar is dissolved and the mixture is almost boiling. Remove from the heat. Immerse the perilla leaves and let cool completely. Strain into a food-safe container and store in the fridge for up to 2 weeks.

The spritz conjures up thoughts of basking in the late-afternoon Italian summer sun while you sip from a giant wineglass brimming with a gorgeous, orange-tinged fizz that'll soon match the color of the sunset kissing the waves of the Mediterranean. Re-create that warm lazy feeling with this soju spritz by layering our Soju Aperitif with bright, bursting bubbles for easy drinking on a sun-blissed summer afternoon or a chill night in.

Fill a wineglass with ice and add the soju aperitif. Top with the sparkling wine and club soda. Garnish with the grapefruit wheel.

SOJU SPRITZ
소주 스프릿츠

Makes 1 cocktail

Ice

2 ounces Soju Aperitif (page 50)

1½ ounces sparkling wine

1½ ounces club soda

Grapefruit wheel, for garnish

YUJA MICHELADA
유자 미첼라다

Makes 6 cocktails

2 ounces yuja (yuzu) juice

2 ounces fresh lime juice

3 ounces pomegranate juice

2 ounces (¼ cup) gochujang

¾ ounce Maggi seasoning

½ ounce soy sauce

6 light beers, such as Hite or Terra

Ice

The michelada, a Mexican drink made with beer, lime juice, and chili spices and sauces, is supremely refreshing come summer. Nick first developed a version of this michelada for our wedding (our signature cocktail list was six deep, befitting a wedding of future bar owners). This one delivers all the punches: sweet and tart from the yuja and pomegranate, spice from the gochujang, and extra umami from the Maggi and the soy sauce.

Drink this on the beach, alongside some fatty samgyeopsal BBQ, or as a refreshing way to combat your hangover.

1. In a blender, combine the yuja juice, lime juice, pomegranate juice, gochujang, Maggi, and soy sauce and blend until smooth. (You can also mix it by hand, but the gochujang might stay a little chunky.)

2. Pour each beer into an ice-filled pint glass, top with 1½ ounces michelada mix, and give it a light stir. Add more beer as you drink.

Maesil-cheong 매실청 (green plum extract) has a subtle floral sweetness that Korean moms and halmeonis reach for when marinating meats or making a tea to soothe their loved ones' upset tummies. It's not something my mom ever really used, so my first interaction with it was when I made my own maesil-cheong at home. I fermented a seasonal pack of roly-poly green plums (acquired from H Mart) in a bed of sugar that dissolved into sweet syrup over the course of the following months.

I like to seek out good maesil-cheong to use in this cocktail— since there are so few ingredients in the drink, you want each to shine. But it will work equally well with any supermarket variety or your grandmother's homemade maesil-cheong. You can adjust the sweetness to taste based on what you end up using.

In a shaker or mixing glass, combine the rye, maesil-cheong, and Angostura bitters. Fill the shaker with ice and stir until well chilled. Strain into a rocks glass over one large ice cube. Squeeze the orange peel over the glass to express the oils, then rub the peel along the inside rim of the glass. You can add the peel as garnish or discard.

MAESIL OLD-FASHIONED
매실 올드 패션드

Makes 1 cocktail

2 ounces rye whiskey
(I'm partial to Old Overholt)

½ ounce maesil-cheong (green plum extract), page 42, or store-bought

2 dashes Angostura bitters

Ice

Orange peel, for garnish

FALL STORY (A WHISKEY SOUR)
가을동화 위스키 사워

Makes 1 cocktail

Ice

2 ounces rye whiskey

¾ ounce Misugaru Syrup
(recipe follows)

¾ ounce lemon juice

Dash of Angostura bitters

Misugaru Syrup

1 cup hot water

4 tablespoons misugaru
(roasted grain powder)

1 cup demerara sugar

The first cocktail I ever loved was an amaretto sour—I think someone made it for me at a faux-fancy college party—which awakened me to the possibility that alcohol could taste good, unlike the watery keg beers and trash-can jungle juices and shots of Bacardi 151 (ugh) I had suffered through until then. I tried in vain to re-create this drink in my dorm, but, not knowing much about anything at the time, I attempted to make it with premade sour mix from the Fresh Grocer, which ended up a sugary abomination. Later, I learned that a proper sour should balance fresh citrus with an interesting sweetener and a high-quality liquor, and I eventually graduated to loving whiskey sours. It's still the drink I turn to when introducing someone new to cocktails.

I named this cocktail after an extremely popular Korean drama (also known as Autumn in My Heart*) that I watched with my mom when I was in middle school; she would rent the tapes from the video store next to the Korean supermarket we visited weekly in the Valley. The confounding thing about this cocktail is that it doesn't taste like its components. Instead, it tastes like a cozy apple cider, which I still don't really understand because there's no apple in the drink. Here, the misugaru, a roasted grain powder, replaces the airiness that is traditionally brought by egg white, adding a warm, comforting sweetness, like a good K-drama.*

In an ice-filled cocktail shaker, combine the rye whiskey, misugaru syrup, and lemon juice and shake vigorously for 20 to 30 seconds to blend. Strain into an ice-filled highball glass and top with a dash of Angostura bitters.

Misugaru Syrup • Makes 1½ cups
In a small saucepan, combine the hot water and misugaru and whisk until completely dissolved. Set over medium heat and add the demerara sugar, stirring until the sugar is dissolved and the mixture is almost boiling. Remove from the heat. Strain into a food-safe container and allow to cool completely; store in the fridge for up to 2 weeks.

The Negroni is my favorite cocktail to order at a bar. It's spirit-forward, making me an efficient drunk, and the proportions—classically equal parts gin, Campari, and vermouth—are easy to remember and therefore a reliable order at most places. This version plays up the sweet and bitter notes of the Negroni with jujube-infused soju, jujube tea, and ginseng. It may seem odd to shake uncooked rice into the drink, but the starchiness will impart a lightly soft texture similar to using ssalttemul 쌀뜨물, or rice water, in a jjigae (stew).

In a cocktail shaker, combine the jujube-infused soju, Campari, sweet vermouth, honey jujube tea, ginseng tea powder, and uncooked rice. Add ice and shake vigorously until well chilled and frothy. Fine-strain into a rocks glass with one large ice cube. Squeeze the orange peel over the glass to express the oils, then rub the peel along the inside rim of the glass and discard.

Notes: The honey jujube tea used here comes in a jar and looks like jam or preserves. It's not a bagged, dried, or powdered tea.

Ginseng tea powder is sold in most Asian grocery stores or online, and comes in individual serving packets meant to be mixed with hot water.

JUJUBE GINSENG NEGRONI
대추 인삼 네그로니

Makes 1 cocktail

2 ounces Jujube Ju (page 46)

½ ounce Campari

½ ounce sweet vermouth

¼ ounce honey jujube tea
(see Notes)

1 (3g) packet ginseng tea powder
(see Notes)

Handful of uncooked white rice

Ice

Orange peel, for garnish

SMOKY BURDOCK MANHATTAN
스모키 우엉 맨해튼

Makes 1 cocktail

Ice

2 ounces Burdock Bourbon
(recipe follows)

1 ounce Lapsang Punt e Mes
(recipe follows)

Burdock Bourbon

12 ounces bourbon

¼ cup dried burdock

Lapsang Punt e Mes

6 ounces Punt e Mes

2 tablespoons lapsang souchong tea

This Manhattan marries the earthiness of burdock—a root vegetable that can be stir-fried as a banchan and used in kimbap or bibimbap, or dried and steeped in tea—with the smokiness of lapsang souchong, a smoked Chinese black tea.

It's based on a classic Manhattan's 2:1 ratio, with 2 ounces whiskey to 1 ounce sweet vermouth. But instead of vermouth I use Punt e Mes, an aperitivo made from vermouth, which contains one point sweetness and a half point of bitterness (quite the golden ratio). If need be, you can substitute sweet vermouth.

In an ice-filled mixing glass, combine the burdock bourbon and lapsang Punt e Mes and stir to chill. Strain into a chilled coupe.

Burdock Bourbon • **Makes 1½ cups**
In a clean jar or glass measuring cup, combine the bourbon and burdock and infuse for one hour. Strain and store at room temperature until ready to use.

Lapsang Punt e Mes • **Makes ¾ cup**
In a clean jar or glass measuring cup, combine the Punt e Mes and tea and infuse for 5 minutes. (Set a timer! Otherwise the infusion will become too tannic.) Strain and store in the fridge until ready to use.

It's taken me a long, long time to appreciate martinis. I used to order them dirty because I thought the addition of olive brine would make me like them more, but then I realized that I hated olives, and therefore hated olive brine. Martinis are such a simple cocktail; like a little black dress, every errant thread or speck of dust will show. Even a drop too much of one ingredient, or choosing the wrong kind of vermouth, can throw the whole drink out of balance. But dasima, the dried kelp that's one of the key umami bases of Korean broths, infuses this martini with a beautifully subtle brininess that keeps me reaching for another. Still, tread lightly, since dasima can get slippery if overinfused in both broth and in alcohol, so take care not to steep it and forget it.

In an ice-filled mixing glass, combine the dasima soju and vermouth and stir to chill. Strain into a chilled coupe.

Dasima Soju • Makes 1 cup
In a clean jar or glass measuring cup, combine the soju, dasima, shiitake, and lemon peel and infuse for 2 to 3 hours (no more!). Strain and store at room temperature until ready to use.

DASIMA MARTINI
다시마 마티니

Makes 1 cocktail

Ice

2½ ounces Dasima Soju
(recipe follows)

¾ ounce vermouth blanc

Dasima Soju

8 ounces Tokki Soju
(or similar 40% ABV soju)

1 (6-inch) piece dasima
(dried kelp)/kombu

1 whole dried shiitake mushroom

A lemon peel

"COLD BREW" LEFTOVER DASIMA BROTH

When you are cooking with dasima (kombu, or kelp), it's important to remove it from the broth before the liquid starts boiling, or the kelp will impart a slippery mouthfeel. But I always feel so wasteful just tossing it. I could slice it up and make a dasima banchan with it, but when I can't be bothered (most of the time), I'll stash it in a quart container and pour cold water over it, making a second, albeit lighter, broth that I can store in the fridge. It's a technique I picked up from Sonoko Sakai's *Japanese Home Cooking* and has the added bonus of ensuring that I will usually have a starter broth on hand, which saves me a ton of time when making dinner. Use it as a base for a quick noodle soup, or for Jogaetang (Spicy Clam Soup) (page 94) or Kongnamul Guk (page 211).

TOKKI TOMATO 3.0

토끼 토마토 3.0

Makes 1 cocktail

Ice

1½ ounces Tokki Soju Black
(or similar 40% ABV soju)

2 ounces Tomato Water
(recipe follows)

1 ounce grapefruit juice, strained

1⅓ ounces Dill Syrup
(recipe follows; see Note)

1⅓ ounces fresh lemon juice
(see Note)

Grapefruit twist, for garnish

Tomato Water

10 medium beefsteak or plum tomatoes

Kosher salt

Dill Syrup

¼ cup fresh dill fronds
(stems discarded)

½ cup sugar

I visited Tōkki Bar on the fourth floor of the trendy RYSE Hotel in Hongdae soon after they opened (and many times after), a beautifully dark and elegant space lit from above by a (faux) skylight, a reverent altar at which to sip their boozy offerings. Their menu simply listed ingredients, like "Burdock" or "White Chocolate," sometimes with numbers that denoted the version.

Tomato 3.0 caught my eye, and I was even more surprised when a highball brimming with a clear liquid was placed in front of me, rather than the red drink I had imagined. The tomato juice that had been clarified, meaning the solids had been removed, and the cocktail was pure umami, with just a little bit of salt and sweetness as balancing notes. My eyes were wide as I immediately downed the drink and ordered another.

While this drink isn't a classic cocktail (yet!), I feel like it's representative of the new wave of American-style cocktails that are diving in to savory flavors and experimenting with clarification. Originally made at their bar with Tokki's Sŏnbi Gin, this recipe has been adapted to feature Tokki Soju instead.

In an ice-filled cocktail shaker, combine the soju, tomato water, grapefruit juice, dill syrup, and lemon juice and shake vigorously until well chilled. Strain into a highball glass filled with large ice cubes. Garnish with a grapefruit twist.

Note: Most jiggers don't have a ⅓-ounce measure; that's okay! Just aim for the space between the ¼ and the ½ mark.

Tomato Water • Makes 1½ cups
Wash the tomatoes and use a sharp paring knife to remove the stems and attached core. Juice or blend the tomatoes and stir in a pinch of salt. Line a sieve with a paper coffee filter or cheesecloth and set over a bowl. Pour the tomato puree into the sieve to drain. Reserve the liquid, storing in the fridge until ready to use.

Dill Syrup • Makes ¾ cup
1. In a blender, combine the dill fronds and ½ cup water and puree. Strain through a coffee filter or a fine-mesh sieve into a small saucepan.
2. Add the sugar to the strained liquid and heat over medium heat until just before boiling, stirring to dissolve the sugar. Turn off the heat and let cool completely. Store in the fridge until ready to use.

CHASING
TASTE
MEMORIES

Every May when the last bell on the last day of school rang, I dragged my feet leaving the campus, trying to get one last signature in my yearbook, sharing another long hug with my friends. For the others, summer meant camp or long days by the pool, but for me summer meant three long sweltering months in Seoul, visiting our grandparents and extended family.

Then, every August, I'd feel the same sad dread when leaving Korea. By then, I'd have had countless bowls of cold noodles with my grandfather, tons of street food snacks in between shopping trips, my eyes and stomach opened by new flavors and tastes and dishes that I would never be able to get back in America. I missed waking up to my halmeoni's breakfast table, heavy with banchan, and spending the days with my cousins.

That nostalgic loss was something my parents must've felt year-round as Korean immigrants in America. They wrangled two young children across oceans every year because they, too, were chasing memories of their youth spent in Korea. Back home in California, my parents would drive us long distances to downtown Koreatown every weekend just to have dinner at a little old-school soondubu (soft tofu stew) spot or a bossam (steamed pork belly) restaurant. Afterward, we would linger in various Korean grocery stores as they endeavored to capture the nostalgia of something they missed from the motherland—a dish, a snack, an ingredient, an ice cream bar. My parents wanted to relive and satisfy their comfort cravings, and in turn, their taste memories filtered into new ones for us.

The cocktails that follow are influenced by my favorite taste memories from those summers spent in Korea as well as my childhood growing up in my Korean American household, ones shared by many who grew up like me, the ones I'm still chasing today.

Whenever I travel to Korea, my go-to convenience store drink is boricha, or barley tea. It feels so much more hydrating than plain water (and for a similar price, anyway). I'll also grab one of those famous coffee pouches that come with an ice cup. After I suck down some coffee, I'll pour in some boricha, creating this new flavor combo in which the roasted nuttiness of the barley tea balances out the bitterness of the coffee. It's my trick for beating jet lag, staying caffeinated and awake while also keeping hydrated for long days traipsing around the city.

I like to use a mix of barley and cassia tea to get that lighter flavor that you find in bottled boricha. For me, barley tea alone has too much of a bitter, bready taste, so I find that mixing in cassia seeds gives it a more rounded sweetness that feels classically Korean. This is also a great hangover drink or a get-the-night-started drink, especially when mixed with a little soju.

Mix all the ingredients and pour over ice.

Boricha • Makes 2½ quarts

1. In a large saucepan, heat 3 quarts water to just before boiling. Reduce the heat to low, add the roasted barley and cassia seeds, and steep for 10 minutes.

2. Remove from the heat and cool completely before straining and transferring the liquid to a lidded pitcher and storing in the fridge. Discard the solids.

BORICHA & COFFEE
보리차와 커피

Makes 1 drink

4 ounces brewed coffee,
at room temperature

2 ounces Boricha (recipe follows)

2 ounces soju (optional)

Boricha

3 tablespoons roasted barley (boricha)

2 tablespoons cassia seeds 결명자

MISUGARU
미숫가루 밀크 펀치

Makes 1 cocktail

4 ounces whole milk

2 tablespoons misugaru
(roasted grain powder)

1 ounce Honey Syrup (recipe follows)

1 ounce dark rum

¾ ounce whiskey
(like Crown Royal) or brandy

Ice

Rice Krispies, for garnish

Honey Syrup

½ cup honey

My dad used to commute all the way to the other side of Los Angeles from where we lived. To avoid a two-hour drive, he would go to work at 6 a.m. and come home by 3 p.m., waking up before the sun to hit the freeways while they were still clear. My mom would wake up even earlier to make him misugaru, a roasted grain powder mixed into milk for a hearty morning drink, a kind of rustic Korean protein shake.

This milk punch is almost like eggnog in its creamy, nearly chewy consistency. Dark rum rounds out the cocktail's body, while the whiskey is a great excuse to break into that bottle of Crown Royal still sitting in its purple velvet sheath on my dad's shelf (why do Korean dads love gifting that bottle so much??). And topping it with a few Rice Krispies gives just the right amount of whimsical pop and crunch.

In a cocktail shaker, combine the milk and misugaru using a handheld blender or milk frother. Add the honey syrup, rum, whiskey or brandy, and ice and shake vigorously to blend. Strain into a rocks glass and top with a few Rice Krispies.

Honey Syrup • Makes ¾ cup
In a small saucepan, mix honey and ½ cup water. Heat until just before boiling, stirring to dissolve the honey. Remove from the heat and let cool completely.

WORKING WITH MISUGARU 미숫가루
The thing I love to hate about misugaru is that it is incredibly hydrophobic, meaning that it is really resistant to dissolving into liquid. No matter how much you stir or shake, there are always clumpy powdery chunks that float on the top, stick to the side, or coat the bottom in a disappointing puddle.

Some tips when making misugaru at home:

Mix with room temperature water, then add ice.

Steep it in milk overnight; give it a few shakes every hour or so before retiring to bed.

Use a hand blender or milk frother.

Binggrae's Banana-Flavored Milk is a childhood icon that's been loved for generations, both for its taste and for the shape of its chubby plastic bottle, reminiscent of traditional Korean jars (known as onggi) and perfect for tiny little hands to wrap around (so cute!). When it was originally introduced in 1974, the Korean government wanted people to drink more milk, then an unfamiliar beverage. Binggrae incorporated banana, a luxury ingredient at the time, into its milk, and the resulting product was so hugely popular that the company followed up with other flavored milks like strawberry, coffee, and melon.

Nick first came up with this cocktail for a Film Feast we hosted at Nitehawk Cinema for the movie Parasite, *inspired by a scene in which the housekeeper's husband, Geun-se, after being trapped without food for days in the basement, hurriedly scarfs down a banana and milk like a little child. The execution of the cocktail is just as simple. Don't skip the Milkis, though; the carbonated sweet yogurt soda coaxes out even more of the banana flavor.*

BANANA MILK
바나나 우유 막걸리

Serves 2

1 (750 ml) bottle makgeolli, such as Màkku

1 ripe banana, cut into chunks

Ice

Milkis

1. In a blender, combine the makgeolli and banana and blend until smooth. Strain through a fine-mesh sieve into a pitcher.

2. Pour into two ice-filled highball glasses until each is about two-thirds full, and top with Milkis. Kick back and cheers with your inner child.

SPAM
스팸 칵테일

Makes 1 cocktail

Ice

1 ounce white rum

1 ounce pineapple juice

½ ounce fresh lime juice

½ ounce Gochugaru Rice Syrup (recipe follows)

¼ ounce Aperol

1 drop liquid smoke

Cubes of fried Spam, for garnish

Gochugaru Rice Syrup

¼ cup rice oligodang (malted rice syrup)

½ teaspoon gochugaru

This drink is a tiki cocktail inspired by Spam with flavors reminiscent of the much-maligned pizza topping: ham and pineapple (which is actually quite beloved in Korea). It's not supposed to taste like Spam (though, trust me, I tried), but instead nods to the spices and smokiness found in our favorite luncheon meat with the gochugaru rice syrup, which adds a deep, rich, malty flavor.

In an ice-filled cocktail shaker, combine the rum, pineapple juice, lime juice, rice syrup, Aperol, and liquid smoke and shake vigorously until well chilled. Strain into a clean Spam can or a rocks glass filled with ice. Garnish with cubes of fried Spam on a skewer.

Gochugaru Rice Syrup • Makes ½ cup

In a small saucepan, combine the oligodang and ¼ cup water and heat until just before boiling, stirring to dissolve. Turn off the heat and stir in the gochugaru. Let cool completely, then store in the fridge until ready to use.

My holy grail dish, the one I'm always chasing whenever I travel back to Seoul, is naengmyeon. Since we always visited in the summer, when the humidity soared and air conditioning was hard to come by, we'd frequent North Korean cold-noodle joints throughout the city with my grandparents. Slurping the chewy noodles as a family while we cooled off together was a core memory that I cherish to this day.

My goal with this recipe is to re-create the flavors and textures that I love about naengmyeon. The brine from dongchimi, a refreshing watery radish kimchi, adds a clean acidity, while the buckwheat tea adds a mouthfeel that's similar to that of the noodles traditionally found in the dish. (Some naengmyeon houses even serve the boiled noodle water as a warm pre-meal tea so you can taste the quality of their buckwheat.)

In an ice-filled cocktail shaker, combine the soju, vermouth, Dasida, brine, buckwheat tea, and lemon juice and shake vigorously until well chilled. Strain into a coupe and garnish with the cucumber.

Buckwheat Tea • Makes ¾ cup

If your buckwheat is not already roasted, toast it in a small saucepan over medium heat until it turns a medium-brown color and smells fragrant. Add ¾ cup water and turn off the heat. Steep until cool, then strain and discard the solids. Store in the fridge until ready to use, no longer than 2 to 3 days.

NAENGMYEON
냉면 칵테일

Makes 1 cocktail

Ice

1¾ ounces high-proof soju (40% ABV)

¼ ounce dry vermouth

⅛ teaspoon Dasida soup stock powder (beef flavor)

½ ounce brine from dongchimi (radish water kimchi)

½ ounce Buckwheat Tea (recipe follows)

¼ ounce lemon juice

Cucumber slice, for garnish

Buckwheat Tea

¼ cup husked buckwheat

DASIDA

Dasida is a soup stock powder that's been around since 1975, one of those magic powders that dissolve in water for an instant broth or make a soup or stew pop with just a few dashes. Don't tell, but it's the secret to many restaurants' dishes, infusing them with that lip-smacking umami without adding too much saltiness.

YAKULT
야쿠르트 소주

Makes 1 cocktail

1 (2.7-ounce) bottle Yakult

2.7 ounces soju (see Note)

2.7 ounces lemon-lime soda,
such as Sprite or 7 Up, or Chilsung Cider

Here are all the reasons I love Yakult, the little red foil-lidded yogurt drink:

1. Because when I was a wee child sweating it out on the summer streets of Seoul, I could always convince my mom to spend the 50 won to buy me one from the Yakult lady.

2. Because I like when it's served frozen at the end of a meal and I get to spend the next 20 minutes attempting to melt mine into the perfect slush; using my hot breath, my tongue, a chopstick—anything but patience or time.

3. Because it's the foundation of Lara Jean and Peter's love story in To All the Boys I've Loved Before *(an iconic movie scene that triggered Yakult nostalgia for Asians everywhere).*

This Yakult soju is a classic Korean combo riding that razor's edge between almost-too-sweet-and-can't-taste-the-alcohol to whoops-I'm-drunk.

Pour the Yakult into an ice-filled glass. Use the empty Yakult bottle to measure the soju and add it to the glass. Do the same with the lemon-lime soda. Mix and serve.

Note: To get the 2.7 ounces called for, use the Yakult bottle to measure the equal amounts.

This recipe comes via Hana Makgeolli, a brewery based in Greenpoint, Brooklyn. Cofounder Alice Jun had been brewing makgeolli at home for years before she opened her own brewery and tasting room, which features makgeolli flights, delicious cocktails like this one, and pitch-perfect anju pairings. She's even since distilled her own soju, made from her makgeolli, as well.

Sujeonggwa is a Korean cinnamon punch that's served as an after-dinner digestive, made with dried persimmon, cinnamon, and ginger. Here, dried jujubes sub in for the persimmon for a more herbal tone that coaxes out the delicate floral notes of Hana Makgeolli's Hwaju 12, which is infused with chrysanthemum flowers and hydrangea tea leaves.

Add all the ingredients to a mixing glass filled with ice. Stir and strain into a chilled coupe. Garnish with some dried persimmon or pine nuts.

Sujeonggwa Syrup • Makes ¾ cup

In a small saucepan, combine the cinnamon stick, ginger, and 1 cup water, Heat until the water just begins to steam and smell fragrant. Turn off the heat and stir in the sugar until dissolved. Let the mixture cool, then strain and store the syrup in the fridge until ready to use.

Jujube Tea • Makes ¾ cup

1. In a small pot, combine the jujubes and 1 cup water and bring to a boil. Reduce to low heat and simmer for 10 minutes.
2. Use the back of a spoon or a muddler to press down on the jujubes, popping them open and crushing them. Cook until the liquid is reduced by half, about 20 minutes. Strain and let cool. Store in the fridge until ready to use.

HANA MAKGEOLLI SUJEONGGWA
하나 막걸리 수정과

Makes 1 cocktail

Ice

2 ounces Hana Makgeolli Hwaju 12, or other light-bodied makgeolli of your choice

½ ounce Sujeonggwa Syrup (recipe follows)

½ ounce Jujube Tea (recipe follows)

3 to 5 dashes Angostura bitters

Dried persimmon or pine nuts, for garnish

Sujeonggwa Syrup

1 cinnamon stick, rinsed

2 tablespoons thinly sliced peeled fresh ginger

1 cup sugar

Jujube Tea

¼ cup dried jujubes

HANA MAKGEOLLI

Hana Makgeolli's offerings rotate often depending on what's in season and whomever Alice is collaborating with, but the standout stalwarts include Takju 16, her signature brew; Omija Makgeolli, infused with the tangy omija; and Yakju 14, a five-time fermented and clarified makgeolli.

MELONA
메로나 막걸리

Makes 1 cocktail

Ice

2 ounces makgeolli, such as Màkku

1 ounce Midori melon liqueur

¾ ounce fresh lime juice

½ ounce simple syrup

Dash of Angostura bitters

When I worked for Màkku, an American-based makgeolli brand created by founder Carol Pak, part of my job was helping market makgeolli to a new generation of drinkers by showcasing how the product could be used in fun cocktails, like this bright green drink that's both eye-catching and nostalgic. It's inspired by Melona, a honeydew melon–flavored ice bar that's been popular in Korea (and among Korean Americans) for decades. At Orion Bar, we serve it as a frosty frozen cocktail, but I like this shaken version when at home.

In an ice-filled cocktail shaker, combine the makgeolli, Midori, lime juice, and simple syrup and shake gently (the makgeolli may fizz and expand). Strain into a coupe glass. Top with a dash of Angostura bitters.

4-CHA

SOOL
PARTY
술 파티

HOW TO HAVE A SOOL PARTY

When the city that never sleeps shuts down, its residents will find a way to keep the party going. Such was the case during Hurricane Irene, a tropical storm anticipated to be so destructive that the New York City mayor closed all businesses and transportation for the entirety of that sweaty August weekend. With nothing to do and nowhere to go, I decided to throw a "Hurricane Me" party at my apartment (the storm and I shared the same name) and invited the friends who could reach me on foot or by bike to hunker down and weather the storm together. On the menu were my favorite things to eat and drink, a perfect opportunity to share my personal Korean American–style comfort foods. While we watched live news updates (the Weather Channel had never been so popular), I first rolled out kimchi-slaw fish tacos and beers, and then fired up my portable gas stove and threw shingles of frozen pork belly and beef brisket on top of a metal Korean BBQ plate. Everyone gathered around to help flip the meat, stuffing wads of lettuce wraps into their cheeks and washing it down with soju. As the storm petered out and passed without much ado, I plonked down a Dutch oven of budae jjigae–style ddukbokki, which we forked into our mouths even as our eyes grew heavy with sleep.

A sool party is literally just a party with alcohol, but the best Korean sool parties are hosted at home like this—more of a never-ending friend hang than a packed blowout party. The goal is to share time together, catch up on life, and ultimately just bond over good food and drink. I like to plan mine around a fun interactive theme or dish: hosting a mini-kimjang where we make kimchi for people to take home and eat alongside freshly boiled bossam and shots of soju; a makgeolli tasting party comparing different brands from the supermarket; or Korean fried chicken from our favorite local takeout spot (hi, Peeps Kitchen!). But it doesn't always need to be so elaborate—you can also just grab a couple bottles of your favorite booze and some snacks. If you've got good company, you've got yourself a sool party.

SOME TIPS & IDEAS FOR HOSTING YOUR OWN **SOOL PARTY**

Stock up on frozen Korean BBQ meats and accoutrements (see Samgyeopsal, page 157).

Plan for one bottle of soju per person (more if your group consists of heavy drinkers) and grab a case of beer and some sodas to supplement.

Create a captivating centerpiece to wow your guests, like a large-format punch served in a whole watermelon (Watermelon Soju Hwachae, page 151) or a big pot of Budae Jjigae (page 169) brimming with Spam morsels and bricks of ramyun noodles. This is definitely the time to try that viral dish or drink you saw online.

Or **enlist your friends** to bring an alcohol and an anju each, potluck-style (and take some of the pressure off yourself).

Course it out to keep the night going instead of serving everything all at once; it will prevent people from getting full and sleepy too soon.

Always **keep a stash of instant ramyun** or late-night snacks; you're bound to get hungry again later.

When you run out of food and conversation, **wind down the night** by firing up the karaoke machine, dancing along to K-pop videos, or turning on an episode of the latest hot drama or a good movie.

When I was growing up, summer was always synonymous with watermelon. My mom and I would take turns carrying the heaviest, ripest watermelon up the hill to my aunt's house in Seoul, forearms on fire by the time we reached the top. To this day, a big juicy watermelon beckoning me outside the grocery store is a challenge I'm always willing to undertake.

This punch is my take on watermelon soju (which is exactly what it sounds like) combined with the fizzy milky fruit dessert called watermelon hwachae. Scoop the watermelon in advance and mix the punch before serving. It'll taste better as it sits and the ice melts, so it's perfect for imbibing at a rooftop party or over the course of many rounds of Korean BBQ.

1. Cut off and discard the top quarter of the watermelon. Position the watermelon in a shallow bowl to keep it steady and upright. Use a melon baller, ice cream scoop, or large spoon to scoop out about 2 cups' worth of watermelon balls (you can also use cookie cutters to make pretty shapes) and set aside. Scoop out the rest of the flesh into a blender until the watermelon is hollow.

2. Blend the scooped watermelon until smooth, and strain through a fine-mesh sieve set over a bowl. Add 2 cups of the watermelon juice to a large pitcher or bowl (save any leftover watermelon juice for another use). Stir in the cider, soju, Campari, lemon juice, condensed milk, and salt until well combined. While stirring, add the milk so it doesn't curdle.

3. Add ice, frozen blueberries, and the reserved watermelon balls to the hollowed-out watermelon. Fill with the watermelon soju and garnish with mint.

4. Ladle into ice-filled glasses to serve.

WATERMELON SOJU HWACHAE
수박 소주 화채

Serves 4 to 6

1 mini watermelon (about 8 pounds)

1½ cups Korean cider, or lemon-lime soda like Sprite or 7Up

1 cup soju

¼ cup Campari

¼ cup fresh lemon juice

2 tablespoons condensed milk

¼ teaspoon kosher salt

¼ cup whole milk

Ice

1 cup blueberries, frozen

Fresh mint sprigs, for garnish

NAKJI SOMYEON (SPICY STIR-FRIED OCTOPUS WITH THIN NOODLES)

낙지볶음 소면

Serves 2, or 4 as an appetizer

OCTOPUS AND NOODLES

2 nakji (small Korean octopus) or squid (about 1½ pounds total), thawed if frozen

2 tablespoons coarse sea salt (or kosher salt)

4 ounces somyeon (wheat noodles)

SAUCE

2 tablespoons gochujang

2 tablespoons gochugaru

1 tablespoon soy sauce

1 tablespoon sugar

1 tablespoon minced garlic

1 tablespoon soju

1½ teaspoons rice vinegar

2 teaspoons fish sauce

TO FINISH

2 tablespoons vegetable or canola oil

2 to 3 scallions, thinly sliced on the diagonal into ¼-inch lengths

1 medium onion, thinly sliced (about ¼ inch)

1 medium carrot, julienned (about ¼ inch)

½ teaspoon sesame oil

1½ teaspoons toasted sesame seeds, plus more for serving

Red-leaf lettuce leaves

Lemon wedges

The first time I ever had anju as it was meant to be had (with alcohol!) was with my aunt, who's identical twins with my mom but lives in Korea—the most alt-universe surrogate mother. We ducked into a local pocha, and she ordered the nakji-bokkeum (stir-fried octopus), along with a bottle of makgeolli, because why not? (My mom rarely drinks, but my aunt will dabble.)

A platter the length of my forearm appeared, piled high with seafood and veggies slicked in spicy red sauce, adorned with mounds of snow-white noodles and a clear plastic glove. I had never seen nakji-bokkeum served like this. My aunt slid on the glove and began mixing everything together, before transferring it by the fistful to my plate. We cooled the spiciness with slurps of the creamy makgeolli, each bite followed by a sip. Soon the bottle was empty, a significant dent made in the mountain of noodles. I was a little buzzed, very full, and eager to explore more of this anju world.

Nakji in Korea is a specific species of small octopus, different from the larger octopus often found in Italian or American cuisine. I find it sold in the seafood section of my local H Mart or in the freezer section packaged in twos, but if it's proving hard to find you can substitute squid, which will have a similar cooking time.

1. Clean the octopus I like to use kitchen scissors to cut into the body of the octopus and remove the whole sac of innards as well as the eyeballs and the beak inside the center hole where the legs meet. Sprinkle the sea salt over the octopus and rub all over, like handwashing laundry, which will clean the octopus and help make it less fishy and more chewy. Rinse thoroughly with water and set aside.

2. Cook the noodles Bring a large wok or pot full of water to a boil and cook the somyeon for 4 minutes or according to the package directions. Scoop out the noodles, leaving the water in the pot, and rinse under cold water until cool.

3. Set up a bowl of ice and water. Bring the pot of water back up to a boil, adding more water, if needed, and add the octopus for about 30 seconds, until the tentacles curl up. Remove and cool in the ice water. Discard the cooking liquid. →

4. Make the sauce In a large bowl, stir together the gochujang, gochugaru, soy sauce, sugar, garlic, soju, rice vinegar, and fish sauce.

5. Chop or cut the octopus into bite-sized pieces (about 1-inch pieces along the tentacles and smaller pieces along the body) and add to the bowl. Toss to combine and set aside.

6. To finish In the empty wok, heat the vegetable oil over medium-high heat. Add the scallions and stir-fry until they are softened and the oil is fragrant, about 2 minutes. Add the onion and carrot and cook until slightly softened, about 2 minutes.

7. Add the octopus and the sauce to the pan and stir-fry for 3 to 4 minutes to cook the octopus and vegetables with the sauce. Top with the sesame oil and sesame seeds. Stir to combine and remove from the heat.

8. Lay 4 pieces of lettuce evenly along the edge of a large platter. Divide the noodles into 4 bundles (you can wrap the noodles around your fingers to create a neat bundle) and pile one on each piece of lettuce. Transfer the octopus to the middle of the platter, sprinkle with more sesame seeds, and squeeze a lemon wedge over it. Serve with a plastic glove for mixing.

Clockwise from far left:
red-leaf lettuce (page 158),
Korean-style grill pan (page 157),
doenjang jjigae (page 159),
kimchi (page 159),
pork belly (page 158),
white rice (page 158)

Samgyeopsal
삼겹살

We can't talk about drinking soju without talking samgyeopsal (pork belly), the ultimate pairing. I can smell it now: the smoky air hanging thick with porky grease, my eyes stinging a bit through the charcoal. Pork fat crackling as it hits the grill, glasses clinking and sloshing and slamming on the table, the din of tables chattering away broken by the occasional "Ajumma!" "Jeogiyo!" Waiters flit by with watercolor-red curlicues of marbled pork, baskets stacked with lettuce leaves. The tables are covered edge to edge with banchan: bright pungent kimchi, bubbling soybean stews, maybe a little ice cream scoop of mac salad. I smell grilled pork and immediately start salivating for soju.

Korean BBQ has long been a mainstay of Korean food culture: a way for friends and family to gather and celebrate birthdays or job promotions, or just to have a good time. It's a pretty straightforward concept: Cook razor-thin cuts or bite-sized pieces of meat tableside to minimize both cooking time and the distance from cooktop to plate—something that's key for people who love to eat their food piping hot. The spread of banchan—side dishes that fill out a traditional Korean meal—that surrounds the central grill serves as a choose-your-own-adventure map for perfectly embellishing your meats of choice. This inherently communal nature makes the whole experience fun and low stakes, and therefore easy to set up at home.

WHAT TO PREP

You can cook samgyeopsal in a cast-iron or nonstick skillet, or any similar griddle, on your stovetop. Or upgrade your tableside setup with a portable camp stove and butane cans. Fire it up with a Korean-style grill pan—its special grooves keep the meat from sticking and help drain grease. It's also worth sourcing Korean BBQ tongs—I love them not only for grilling but for cooking in general.

WHAT TO BUY

Most Asian grocery stores will have everything you need: short- or medium-grain white rice, fresh red-leaf lettuce and perilla leaves (my personal favorite!), and premade banchan.

THE PORK BELLY

Korean BBQ at home is made easy by the availability of presliced meats at Asian grocery stores. Pork belly comes in two options: thinly sliced or thickly sliced. The thinly sliced pork belly is shaved into curlicues and cooks up super quickly and crisply, so it's great for an impatient group or as a first round. Thick slices are more akin to what you'd find at Korean BBQ restaurants nowadays; they'll cook and sizzle in their own fat as you sip your soju and gab with your table, resulting in juicy, unctuous bites.

When picking your pork belly, look for good marbling: where the white fat is distributed beautifully through the pink-red muscle. The more white speckles the better, and a few striations or strips of white are good, but I would avoid pieces that are a lot more than half fat or feature stringy, tough-looking streaks of fat. Pork belly will classically have three or five stripes of fat—you'll want these to be tightly layered with the stripes of meat.

Plan for one-quarter to one-half pound of meat per person, adjusting for people's appetites and the amount of other dishes you plan to have to fill out the table. A smaller portion can go a long way when wrapped with spoonfuls of rice in filling lettuce wraps, but leftover meat can easily be thrown into stews or ramyun as a round two.

SIDES & SAUCES

This is where the fun begins! The rich, meaty samgyeopsal holds up well to bold flavors, seasonings, and textures like:

Pa-muchim 파무침—shaved scallion salad

Buchu-muchim 부추무침—seasoned garlic chives

Mu-saengchae 무생채—spicy or sweet-and-sour radish salad

Ssamjang 쌈장—a salty, savory paste made with doenjang, gochujang, garlic, sesame oil, and other aromatics, used for dipping meats and making ssam (wraps)

Kimchi

Raw garlic cloves

Red- or green-leaf lettuce or perilla leaves, washed and dried

Rice

Doenjang jjigae or kimchi jjigae

HOW TO COOK

Once your table is ready, heat up your grill on the burner—use a small piece of fat to oil the grill, or a paper towel lightly soaked in vegetable or sesame oil.

Keep sliced meat frozen until cook time—you can take it out 5 to 10 minutes beforehand to help separate the pieces. Lay out each piece on the grill so they don't overlap, and cook on one side until the edges start to curl. Flip and cook until your desired doneness (floppy or extra crispy). Since it's your BBQ, it's your call!

BUILD YOUR SSAM

As the cooked meats get plucked off the grill, prepare your lettuce wrap in one hand (you can rip large leaves in half) and place the meat in the leaf's center. Dollop on a bit of ssamjang and whatever other accoutrements float your boat, like a spoonful of rice to fill out your bite, some pa-muchim to add crunch and acid, or any banchan or kimchi. Then just fold the leaf over your stack and pop the whole thing into your mouth! Repeat and repeat until your heart is content and your belly is full.

*Fried chicken is so iconic as an anju that it spawned its own portmanteau for its go-to drink pairing: **chimaek,** aka **chi**cken and **maek**ju (beer). The combo rose to fame during the 2002 Korea/Japan World Cup Games, when spectators flooded hofs (beer pubs) and ordered pitchers of cold light beer and platters of chicken en masse. Nowadays, Korean fried chicken is so popular it's edging out the Colonel for the title of KFC.*

In the world of Korean fried chicken, there have traditionally been two major camps: soy garlic versus spicy. (Today there are a lot more camps: buldak, mala, sprinkle, or snowing cheese, but I digress.) While I personally prefer soy garlic, this recipe meets somewhere in the middle with a flavorful sweet-and-salty-and-just-a-little-spicy sauce.

CHIMAEK CHICKEN
치맥 치킨

Serves 4

CHICKEN

2 pounds chicken drumettes

1 cup kimchi brine (optional)

2 teaspoons kosher salt

¼ cup potato starch

SAUCE

¼ cup soy sauce

¼ cup mirin

2 tablespoons minced garlic

2 tablespoons honey

2 tablespoons sugar

2 teaspoons gochugaru

1 teaspoon toasted sesame seeds

1 teaspoon mustard powder

¼ teaspoon ground white pepper

1 scallion, chopped

1 tablespoon cornstarch

TO COOK

Vegetable or canola oil, for frying

1. Prepare the chicken In a large bowl, toss the chicken drumettes with the kimchi brine and refrigerate for 1 hour, uncovered. This will help tenderize the meat and is a great use for leftover kimchi juice, but you can skip if you don't have any on hand.

2. Drain the chicken and discard the kimchi brine. Sprinkle the chicken with the salt and then with the potato starch. Toss to coat.

3. Make the sauce In a small saucepan, combine soy sauce, mirin, garlic, honey, sugar, gochugaru, sesame seeds, mustard powder, white pepper, and scallion. Set over low heat until the sauce just begins to boil.

4. Meanwhile, in a small bowl, combine the cornstarch and 2 tablespoons water. When the sauce just begins to bubble, stir in the cornstarch mixture and simmer until the sauce is well blended and glossy, 2 to 3 minutes. Turn off the heat and set aside.

5. To cook When you are ready to fry the chicken, line a large plate or bowl with paper towels and keep it near the stove. Pour enough oil into a medium saucepan or a large wok to submerge a single layer of wings. Heat the oil over medium heat to 300°F.

6. Working in batches so as not to crowd the pan, add the chicken and fry for 5 minutes. Transfer to the paper towels. Increase the heat to medium-high and bring the oil to 375°F.

7. Fry the chicken again for another 5 minutes. Drain on the paper towels. →

8. Transfer the chicken to a metal bowl, pour in the sauce, and toss the chicken in the sauce to coat. Serve immediately with lots of ice-cold beer.

Note: This recipe calls for both potato starch and cornstarch, the former for coating the chicken before frying and the latter for thickening the sauce before glazing. In a pinch, you can swap one for the other since both starches have an aerating quality when fried (for that nice crispy crunch) and a thickening quality when cooked (for the perfect sauce coating).

When I was a kid, there were only two foods that we could get delivered to my halmeoni's apartment. One was classic Korean-Chinese food, like the still-iconic jjajangmyeon (savory black bean noodles) and tangsuyuk (sweet and sour chicken), and the other was Korean pizza. Halmeoni absolutely delighted in ordering a pie from Mr. Pizza, which looked a lot like the Pizza Hut pizza we would get back in the States but with completely different toppings. Ham and pineapple was beloved, and sometimes sliced hot dogs, but always corn and a ranch or mayo sauce, like a Korean fever dream interpretation of what they thought American pizza should be.

You can use store-bought pizza dough for this recipe, but my secret hack is to go to my favorite pie shop and ask them for a ball of their dough, which will usually be just a few bucks and easily make 2 to 4 pies. The dough recipe below comes from J. Kenji López-Alt's Detroit-Style Pan Pizza, since he's already done the hard work of honing the doughy crust that defines Korean-style pizza.

1. Make the dough In a stand mixer fitted with the dough hook, mix the flour, yeast, and salt until combined. Add just under 1 cup (220g) water. Mix on low speed until the dough comes together into a rough ball, then let rest for 10 minutes. Mix again on medium-low speed until the dough forms a smooth ball, about 10 minutes more. Cover the dough tightly with plastic wrap and set aside in a warm place for about 2 hours, until it has doubled in volume.

2. Make the sauce In a medium wok or skillet, heat the oil over medium heat. Add the onion and salt and cook, stirring occasionally, until the onion just starts to brown, 7 to 8 minutes. Add the garlic, ginger, and scallion and cook for another 30 seconds while stirring. Add the gochujang and stir until incorporated, about another 30 seconds. Add the canned tomatoes (you can rinse the can with a little water and add that to the pot) and stir to incorporate. Reduce the heat to low, add the sugar, and bring the sauce to a low bubble. Simmer for about 30 minutes until the sauce is slightly reduced and remove from the heat. If you wish, you can blend this into a smooth sauce or leave as is for a more rustic, chunky texture.

3. Prepare the corn In a dry nonstick wok or skillet, heat the frozen corn over medium-high heat. Season with the salt and pepper and cook, shaking the pan occasionally, until cooked →

KOREAN PIZZA
코리안 피자

Makes 2 pizzas

DOUGH

2 heaping cups (300g) all-purpose or bread flour

1 teaspoon (5g) instant yeast

1 tablespoon (9g) Diamond Crystal kosher salt

SAUCE

1½ teaspoons oil, like vegetable or canola

¼ onion, diced

¼ teaspoon kosher salt

1½ teaspoons minced garlic

¾ teaspoon minced fresh ginger

1 scallion, chopped

1 tablespoon gochujang

1 (15-ounce) can crushed tomatoes

1½ teaspoons sugar

CORN

¾ cup frozen corn kernels (no need to thaw)

½ teaspoon kosher salt

¼ teaspoon freshly ground black pepper

½ teaspoon gochugaru

BULGOGI

½ pound thinly sliced rib eye (see Note)

2 tablespoons soy sauce

2 teaspoons sesame oil

1½ teaspoons sugar

1½ teaspoons minced garlic

1 scallion, thinly sliced on the diagonal (about ¼ inch) →

through but still toothy, about 3 minutes. Stir in the gochugaru and remove from the heat.

4. Make the bulgogi In a bowl, toss the rib eye with the soy sauce, sesame oil, sugar, garlic, and scallion and marinate for at least 10 minutes.

5. When ready to cook, heat a nonstick wok or skillet over medium heat. Add the bulgogi, stirring frequently and using chopsticks to shred any large pieces, until most of the meat is no longer pink, about 3 minutes. It's fine to undercook a bit since it will continue to cook on top of the pizza. Remove from the heat.

6. Assemble the pizza Position a rack in the center of the oven and preheat the oven to 550°F, or as high as your oven will go.

7. Spread 1½ teaspoons oil in a cast-iron skillet or pizza pan, moisten your fingers with more oil, and transfer half the dough to the pan. Gently spread the dough toward the edges; it will likely spring back but do your best to stretch and loosen it. Let sit for 10 minutes, then repeat (the dough should now be more elastic). Pull the dough up the edges of the pan, being careful not to tear it (if you do, just pinch the dough back into place).

8. Spread ¼ cup sauce in the middle of the pizza, leaving about 1 inch of crust all around the edges. Sprinkle all over with half the Parmesan and then half the mozzarella. Dot the pizza with half the bulgogi, then sprinkle half the corn evenly over top.

9. Transfer to the oven and bake until the bottom of the pizza is a medium golden brown, 14 to 15 minutes.

10. Remove the pan from the oven and remove the pizza from the pan when cool enough to handle. Repeat with the remaining dough, sauce, and toppings. Drizzle the ranch generously on top, and slice.

11. To serve Pass more ranch at the table along with some kimchi or jalapeño pickles—and copious amounts of beer.

Note: You can use premarinated meat from Asian grocery stores; in that case, skip marinating the bulgogi and go straight to cooking it. If you buy whole rib eye, I like to put it in the freezer for about 30 minutes to make it easier to slice, then slice it a little more thickly than you might find at Korean BBQ restaurants, about ¼ inch thick, because the meat will lose moisture during the pizza cooking time.

ASSEMBLY

Olive or canola oil, as needed

½ cup grated Parmesan cheese

2 cups shredded mozzarella cheese

¼ cup ranch dressing

FOR SERVING

Ranch dressing

Kimchi or Soy Sauce
Jalapeño Pickles (page 66)

For many of us, nachos were the first food we ever made for ourselves. Maybe you were a kid in search of a snack to eat while watching TV, or maybe you were out too late drinking with your friends and all the stores had already closed. All you needed was a handful of chips, a fistful of cheese, and a microwave. Sad, but passable.

I discovered the true potential of what nachos really could be while working my first office job out of college. My team were a bunch of food-and-drink-loving engineers, who would start debating lunch options as soon as they arrived at the office. A regular spot for us was a tavern called Souths on Church Street in TriBeCa (alas, no longer). The lunches we had there were my introduction to an American-style hwesik (a group gathering). We'd order a round of black and tan beers (half Guinness, half lager), maybe two, maybe a pitcher along with our lunches, and we'd waddle back to the office afterward, very full and a little buzzed, to work through the rest of the afternoon. But the crowning jewel of a lunch at Souths was their nachos, a plate so towering it became a tradition to welcome any new co-worker by ordering one for the table and seeing if they could pull their own weight.

I wanted to build my own legendary plate of nachos to honor the spirit of all those boozy work lunches, something impressively towering, stacked high to feed a crowd, and infused with punchy Korean flavors. The key to the construction of these nachos is to layer the bulgogi-style pork with copious amounts of cheese throughout the stacks of chips and briefly microwave them before topping it with heaps of sour cream, guacamole, and pickled jalapeños. While you can sub out any ingredients you might not have on hand, definitely don't skip the garlicky scallion cilantro gremolata.

1. **Make the scallion cilantro gremolata** In a small bowl, stir together the scallions, cilantro, olive oil, sesame seeds, garlic, lime zest, lime juice, gochugaru, salt, and pepper.

2. **Make the stir-fried kimchi** In a wok or skillet, heat 1½ teaspoons of oil over medium-high heat, and add the kimchi. Cook, stirring occasionally, until charred on the edges, about 10 minutes, adding the remaining 1½ teaspoons oil about halfway through when the pan feels dry. Remove the →

LOADED DWAEJI BULGOGI NACHOS
돼지불고기 나초

Serves 4 to 6 as an appetizer

SCALLION CILANTRO GREMOLATA

4 scallions, minced

½ cup loosely packed fresh cilantro leaves, chopped

1 tablespoon olive oil

1 tablespoon toasted sesame seeds

1½ teaspoons minced garlic

Grated zest and juice of ¼ lime

½ teaspoon gochugaru

½ teaspoon kosher salt

¼ teaspoon freshly ground black pepper

STIR-FRIED KIMCHI

1 tablespoon neutral oil

½ cup loosely packed kimchi, roughly chopped

BULGOGI GROUND PORK

2¼ teaspoons gochujang

1½ teaspoons doenjang

1½ teaspoons soy sauce

½ teaspoon sesame oil

¾ teaspoon mirin

¾ teaspoon toasted sesame seeds

¾ teaspoon minced garlic

½ pound ground pork

1 tablespoon neutral oil

ASSEMBLY

6 ounces tortilla chips (about ½ bag)

1 cup shredded mozzarella cheese

¼ cup crumbled queso fresco or ricotta salata

½ cup sour cream

1 tablespoon sliced pickled jalapeños, homemade (page 66) or store-bought

kimchi from the pan and use a paper towel to wipe out any burnt bits.

3. Make the bulgogi ground pork In a small bowl, combine the gochujang, doenjang, soy sauce, sesame oil, mirin, sesame seeds, and garlic. Set the sauce aside.

4. Add the oil to the same wok or skillet over medium heat. Add the pork and cook until it is about 70 percent cooked through, 3 to 4 minutes. Add the sauce and cook for another 5 minutes. Remove from the heat.

5. Assemble the nachos! Spread one-third of the chips on a large microwave-safe plate. Top with one-third of the bulgogi ground pork and one-third of the mozzarella. Repeat, this time also adding half the cooked kimchi. Pile on the rest of the chips, pork, and cheese.

6. Microwave for 2 minutes, until the cheese is melted. Top with the rest of the kimchi, and sprinkle on the queso fresco and scallion cilantro gremolata. Dollop the sour cream on top, and scatter the pickled jalapeños around the nachos.

Also known as army stew, this hodgepodge flavor bomb of processed meats, kimchi, and veggies is a dish my family has loved my whole life. My halmeoni was especially partial to it and would take me to a budae jjigae restaurant on the top floor of the Lotte Department Store as a break when shopping (the thought of this fancy restaurant serving Spam and instant noodles still tickles me to this day). Budae jjigae remains my top stew when drinking, for hangovers, or really anytime I get a hankering.

I've had (and made!) many, many budae jjigaes throughout my lifetime, and this is hands-down the best version. It's old-school with all the requisite components, like the budae jjigae I used to have with my grandma, but my secret is to add a dollop of refried beans to the sauce, infusing the stew with a deep rich flavor, before topping it with a tangle of fresh Korean herbs to balance it out with a refreshing brightness. This is also a great way to use leftover broth that you might have knocking around your fridge, or use my "cold brew" anchovy broth (see page 121).

1. Make the sauce In a small bowl, mix together the refried beans, garlic, gochujang, soy sauce, gochugaru, sugar, ramyun flakes, and ramyun powder. Set the sauce aside.

2. Prepare the stew If using hot dogs, cut first into thirds. Use a sharp paring knife to cut "octopus legs" on one end—make a slice from halfway up the hot dog third or weenie down to one end, turn 90 degrees and make another slice, splitting into 4 "legs."

3. Add the prepared sauce to the middle of a wide shallow cooking pot or braiser. Arrange the hot dogs, Spam, tofu, mushrooms, kimchi, rice cakes, and scallions around the pot. Add 3 cups of the broth and bring to a boil over high heat. You can cook this at the table on a portable gas stove, if you have one.

4. Once the pot is boiling, give the stew a little stir to incorporate any residual sauce, and nestle in the ramyun noodles. Top with enoki mushrooms and greens (if using). Reduce the heat to medium, and cook until the noodles are cooked through, another 3 to 4 minutes, Top with cheese (if using). If cooking on a portable gas stove, you can turn the heat down to low to keep the stew warm as you eat, adding more broth (or water) if needed.

5. Serve with warm rice, if desired—and lots of soju.

BUDAE JJIGAE
부대찌개

Serves 3 or 4

SAUCE

½ cup refried beans

1 tablespoon minced garlic

1 tablespoon gochujang

1 tablespoon soy sauce

1 tablespoon gochugaru

1½ teaspoons sugar

Ramyun flakes and seasoning powder, from 1 (4.23-ounce) package ramyun, preferably Shin Ramyun or Samyang Ramen

STEW

2 hot dogs or 6 cocktail weenies

½ can Spam, sliced

¼ container tofu (about 4 ounces), sliced

4 shiitake mushrooms, sliced

½ cup kimchi, roughly chopped

½ cup Korean rice cakes (cylinders or disks)

3 scallions, sliced on the diagonal into 1-inch lengths

3 to 4 cups anchovy broth, veggie broth, or pork bone broth

Noodles from 1 (4.23-ounce) package ramyun, preferably Shin Ramyun or Samyang Ramen

3½ ounces enoki mushrooms, root section trimmed, separated into small bundles

1 bunch minari, crown daisy, or thinly sliced perilla leaves (optional)

1 or 2 slices American cheese (optional)

Cooked rice (optional), for serving

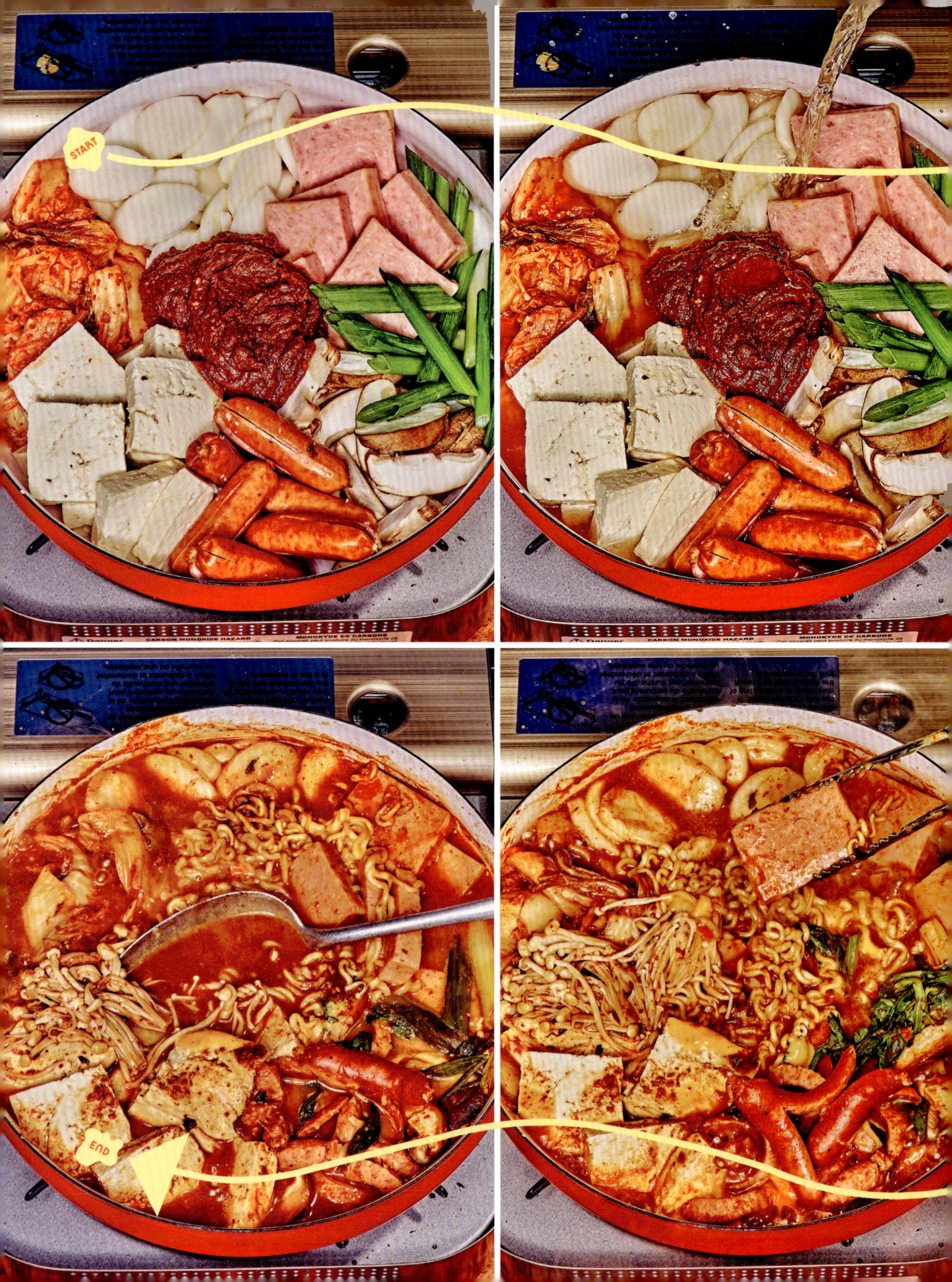
START
END

Sure, drinking games may seem stupid.

What's the point? Shouldn't we just be able to sit and sip our cocktails and have civil conversations like grown adults?

I can see why you might feel that way. Learning how to drink with my friends growing up in America, we played lots of stupid drinking games. I remember my first round of King's Cup at a high school party, playing cards spread around a laughably large brandy snifter, into which we would pour glugs of our various drinks, and whoever pulled the last king had to down the whole thing. In college, my off-campus house was renowned for hosting epic Beer Pong tournaments (I even built and hand-painted our own table). Eventually, we grew up and left drinking games like this behind as we graduated to barstools and bar tabs, cocktail parties and tasteful bottles of wine.

But I don't feel like I'll ever grow out of Korean drinking games. The goal, of course, is the same: get drunk or get someone else drunk. But in Korea they serve as icebreakers in a culture that is not naturally as forthcoming as that of the United States and blur the rigid hierarchy established upon first meeting. Koreans have been playing drinking games since ancient times, when it was commonplace for the losers to have to serve the winners alcohol and food, with everyone ultimately gathering to drink, sing, and dance.

What's unique about Korean drinking games is that they are usually incredibly straightforward in rules and execution, requiring little skill beyond counting, and are easy to pick up as you go. They're so simple that children can play them, and maybe that's why they're still fun as (drunk) adults.

TWIST → FLICK

YOU WIN, NEIGHBORS DRINK!

TWIST CAP
병뚜껑 치기

This is my favorite game to play when drinking with a new group, because it requires very little introduction or skill.

After wowing your compatriots with a perfect soju tornado (see page xlvii) and elaborate bottle opening (see How to Open a Soju Bottle, page xlvi), save the twist-off cap, with the metal ring hanging by a thread from the bottom, like a tail. Carefully twist this piece into a tight strand, and take turns trying to flick it off the cap before passing it to the next person. Whoever successfully flicks the piece off the cap is the winner, and the person on each side of them has to drink.

TITANIC
타이타닉

Fill a glass partway with beer and carefully float an empty soju glass in the liquid. Take turns pouring soju into the glass. Whoever sinks the glass has to drink the whole thing!

SAM-YUK-GU GAME
369 게임 (3-6-9 GAME)

The 3-6-9 game is simple but deadly, especially the more you drink and the fewer faculties you have control over. Take turns going around the table and counting, starting with 1, but on any number that contains a 3, 6, or 9, the person who calls the number must clap. If the numbers appear more than once (e.g., 33 or 36), the person must clap twice to represent both. Whoever claps when they shouldn't or says the wrong number has to drink. The faster you play the game, the harder it gets!

DDALGI GAME
딸기 게임
(STRAWBERRY GAME AKA FRYING PAN GAME)

This is a rhythm game, where the goal is to "toss" to your opponents (hence the alternate name "frying pan game") and try to trip them up. Everyone begins clapping in the same repeating pattern:

1. Clap both hands on the table or your lap.
2. Clap your hands together.
3. Stick your right thumb out to the right.
4. Stick your left thumb out to the left.

Taking turns, the first player (any player can start) calls out another person's name, followed by a number. That player must say "ddalgi" 딸기 (Korean for "strawberry") the requested number of times, on beat, and then call out another player and number. For example, if Jane calls out "John 3," John must say "ddalgi" on the second, third, and fourth beat in time with the above pattern. Whoever misses the beat loses and sits out the next round, and the remaining players begin anew until a winner emerges.

Alternate version: Each player picks a name for themselves, ideally from the same category (for example, animals or maybe a favorite dish), and the chosen names are called out instead of the players' names and the word *ddalgi* for extra layers of complexity.

BUNNY GAME
바니 게임

The first player (any player can start) says "Bunny Bunny" "바니 바니" while making an eating motion with their hands to their mouth, and then says "Bunny Bunny" while making the same gesture to another player. The receiving player must then say "Bunny Bunny" with their hands to their mouth, and the players on each side of the receiving player say "Danggeun Danggeun" "당근 당근" (*danggeun* is Korean for "carrot") while moving their arms up and down, elbows at right angles. The receiving player then passes the Bunny Bunny to another player, who must react as the Bunny with the two people flanking them responding as the carrots. The goal is to play as fast as possible until somebody messes up and has to drink. Got the hang of it? Double up by passing *two* Bunnies at the same time! If the same person gets called on by both bunnies, then they can pass twice to split it up again.

RIDING THE WAVE

파도타기

One person starts drinking, and then the person next to them starts, and so on, like doing the wave in a baseball stadium. The catch is, you can't stop drinking until the person before you stops.

LET'S BUILD AN APARTMENT
SIXTEEN STORIES HIGH!

APARTMENT GAME
아파트 게임

This super simple counting game was repopularized by Rosé and Bruno Mars's 2024 hit song "APT." This game is best played with at least three people (playing with only two people makes it very easy to cheat).

Start seated in a circle so that everyone can reach each other's hands. On someone's count, everyone puts their two hands in the middle, overlapping at random in a human-hand Jenga tower (like an apartment building!). One person calls out a random number. The person whose hand is at the bottom of the stack places it on top, counting "1," followed by the person whose hand is now at the bottom placing it on top, counting "2," and so on until someone reaches the chosen number called. The person with their hand on top is the winner, and everyone else drinks (or, alternatively, that person is the loser, and only they drink).

NUNCHI GAME
눈치 게임

The point of this game is to test one of the most Korean of concepts: nunchi, a social skill that combines innate self-awareness with the ability to read the room. One person begins by counting "1," and everyone takes random turns counting off. The goal is to *not* say a number at the same time as someone else. Whoever says the same number at the same time loses and must drink.

DRINKING GAME ETIQUETTE

For any game, if you or someone in your group gets too drunk to take the penalty shots, you can designate someone to drink in your stead. This person is known as a black knight 흑기사 or a black rose 흑장미 depending on gender. Their reward for swooping in to save your night? You must grant them a wish!

OVER

5-CHA

5차

해장

PARTY'S OVER
(IT AIN'T OVER)

Congratulations, you're hungover!

You did it. You got drunk last night: You spun all the bottles, drank all the soju, ate all the anju, then drank some more.

Korean drinking is not complete until you've resolved the morning after. Hangover management is so important to Koreans that there's both an entire industry *and* a food group to support it; unsurprising, really, since so much of the culture is based on the need to drink heavily any night of the week and make it to work on time the next day. Every convenience store sells a wide variety of hangover remedies right by the counter, in jelly, powder, capsule, or liquid form. Better yet are the hangover restaurants, establishments that operate 24 hours to service those who are still drunk and those who were recently drunk (or really just anyone looking for a hearty meal at any time of day). If you're next-level hungover, then the local pharmacist will hook you up—let them know your exact symptoms (upset stomach? Headache? What did you drink last night?)—and give you a tailored mix of powders and pills to get you just right.

But I secretly relish waking up with a hangover because it's the perfect opportunity to treat it with haejang guk.

Literally "hangover soup," haejang guk is one of a number of foods that Koreans seek out to soothe their queasy bellies and pounding headaches. The term refers to both a specific soup (page 208) and any soup used to get over a hangover, like kongnamul guk (page 211) or even instant ramyun (page 215). Reach for these drinks and dishes the next time you're not feeling your best, and you might find yourself looking forward to the morning after a little more often (same time tomorrow?).

CONVENIENCE STORE HANGOVER DRINKS

Every convenience store throughout Korea is stocked with powder sticks, jellies, and tinctures touting special vitamins and herbal ingredients purported to prevent tomorrow's headspin if taken before drinking or at bedtime. A few standouts:

- **DAWN 808 여명 808:** One of the first hangover drinks introduced in Korea. The name comes from Dr. Jong Hyun Nam's 808 tries to land on this hangover-reducing formula, which contains herbal medicinal ingredients like hazelnut, jujube, ginger, and arrowroot.

- **CONDITION STICK 컨디션 스틱:** These slurpable jelly sticks produced by inno.N are a more modern, increasingly common style of hangover cure, since the sweet candy-like flavor is much more palatable than traditional Asian medicine.

- **EASY TOMORROW 상쾌한:** Each little packet contains a small handful of pills that are meant to be swallowed whole with water. They contain oriental raisin and Chinese hawthorn, which are both popular hangover cure ingredients in Eastern medicine.

I've been waking up to drink Bloody Marys ever since my college friends and I would gather to make big pitchers of them on weekend mornings to pregame brunch or a football game. The saltiness feels restorative while the spicy kick jolts me awake long enough to ease back into a nice little buzz. As the years have gone by, I've been perfecting my recipe—tweaking it just-so to get the right amount of savory umami tartness. The kimchi brine is the secret hero here; just a splash of it brightens the cocktail while deepening it with a little funky je ne sais quoi. Don't believe me? Ask the Bloody Mary G.O.A.T. trophy I won at the Food Network office party!

1. In an ice-filled cocktail shaker, combine the soju, tomato juice, lime juice, lemon juice, soy sauce, kimchi brine, and fish sauce and shake to blend. (Alternatively, use two pint glasses to pour it back and forth to mix.)

2. If desired, mix equal amounts salt and gochugaru on a small plate. Moisten the rim of the glass with a lime wedge or water, then dip the rim in the gochugaru salt and twist to coat.

3. Pour the drink into an ice-filled pint glass. Garnish with some kimchi skewered on a toothpick (if using).

KIMCHI BLOODY MARY
김치 블러디 메리

Makes 1 cocktail

Ice

2 ounces soju

4 ounces tomato juice

¼ ounce fresh lime juice

¼ ounce fresh lemon juice

¼ ounce soy sauce

½ ounce kimchi brine

A few dashes of fish sauce

Kosher salt and gochugaru (optional), for garnish

Kimchi (optional), for garnish

MAKE THIS BLOODY YOUR OWN!

Feeling shy about fish sauce? Sub in Worcestershire sauce instead; it will give you the same acidic umami brininess. You can also sub in vodka, gin, or tequila for the soju. Mezcal is a personal favorite of mine here; it adds a deep smokiness.

An Ode to Pocari Sweat

I meant to create a Pocari Sweat–based hangover cocktail, but the more I thought about it, the more I realized that this Japanese sports drink *is* the hangover cocktail; it's perfect as is. I did not understand this drink as a child—it was one of the three sodas usually available in Korean vending machines, but I much preferred the creamy and fizzy Milkis or the lemony Chilsung Cider over the bland and flat Pocari Sweat. I thought it tasted like salty water, or sweaty soda, like its name.

But as a hard-drinking adult, I have found Pocari Sweat to be my lifesaver in more ways than I can count. Sometimes soda is *too* bubbly and cloyingly sweet. Other hangover cure options, like Gatorade, rely too heavily on sugar and definitely too much on Red 40 food dye. And water is often not enough; no matter how much I drink, I still feel ghastly and thirsty.

The electrolyte-rich Pocari Sweat is meant to replace the nutrients lost when sweating—the same electrolytes that my body is craving after I've been drinking. Since it's not carbonated, it's more chuggable, delivering faster relief of my poor pounding headache. And it's handy in other situations, too: on a particularly hot and draining day, after a taxing workout, or when you have a serious case of the flu and can't keep anything else down.

Nowadays, I'll keep a one-liter bottle stashed in the back of my fridge so I'm always ready to combat hangovers. I need nothing else.

Other hangover cures I swear by:

- A half glass of water with two Alka-Seltzer tablets and a packet of Emergen-C (the former to settle your stomach, the latter to boost your immunity)

- A bacon-egg-and-cheese sandwich from my bodega in Brooklyn (or a ham-egg-and-cheese toast from any Isaac Toast in Seoul)

- A giant carton of coconut water or watermelon juice

- Three Advil and going back to bed

HAEJANG GUK (HANGOVER SOUP)

우거지 해장국

Serves 2

UGEOJI

6 outer napa cabbage leaves

SOUP

1½ teaspoons sesame oil

½ pound beef chuck, thinly sliced

½ cup thinly sliced mu (Korean radish) or daikon, cut into 1-inch squares

1 tablespoon doenjang

1 tablespoon minced garlic

1½ teaspoons soup soy sauce (see page 221)

2 tablespoons gochugaru

4 cups beef bone broth

4 scallions, cut into 2-inch lengths

1 packed cup soybean sprouts, rinsed

½ teaspoon fish sauce

Cooked rice, for serving

My umma's favorite soup is haejang guk (literally, "hangover soup"), which is funny because she never drinks. It still makes me laugh to think back to my petite little mom taking her kids to a hole-in-the-wall haejang guk place for lunch, surrounded by ajeossis sweating out their hangovers around us, blowing their noses and wiping their brows and calling the ajumma over to ask for a bottle of soju, a little hair of the dog.

My mom loves lots of ugeoji (here, dried napa cabbage leaves) in her haejang guk. (She also loves when it has lots of seonji, or coagulated blood, but that's a recipe for another book.) It's kind of a genius food-waste hack to take the unsightly, larger, tougher leaves that surround the tender napa cabbage, dry them, and then reconstitute them for the umami-rich base of a homey soup.

If you can get your hands on some homemade beef bone broth that's so thick, it's Jell-O at room temperature, this is the perfect dish to use it for. You can also use bags of Korean-style beef bone broth, which are great to have on hand (though not quite as thick and jiggly). If you find that your only option is using standard grocery store beef broth, or water, you can "beef it up" with a hunk of beef, beef bone, kelp, or dried mushrooms for additional umami, depth, and hangover-fighting nutrients.

1. Prepare the ugeoji Bring a pot of water to a boil over high heat. Add the napa cabbage leaves and cook for 5 minutes. Remove and rinse in cold water until cool. Cut off any unsightly brown edges from the stem, then shred the leaves into small pieces. Squeeze to drain excess moisture. You can freeze this ugeoji until you're ready to use it.

2. Make the soup In a medium pot or ttukbaegi (a Korean earthenware pot used for boiling soup), heat the sesame oil over medium heat. Add the beef and stir-fry for 1 to 2 minutes, then add the ugeoji. Cook for 4 to 5 minutes, which will help remove any excess moisture from the cabbage. Add the radish, doenjang, garlic, soup soy sauce, and gochugaru and stir to combine. Add the beef bone broth, bring to a boil, and cook for 20 minutes.

3. Add the scallions and soybean sprouts. Reduce the heat to a simmer and cook for another 5 minutes; the ugeoji should be deeply cooked and have taken on the color of the doenjang soup. Finish with the fish sauce and serve immediately, piping hot, with bowls of warm rice and lots of napkins to sop up your hangover sweat.

Do not underestimate this simple soup. Sure, it may look like just sprouts boiled in water, but secretly it's an umami- and nutrient-dense, rejuvenating one-two punch. The kongnamul (soybean sprouts) release so much refreshing flavor, and it's perfectly seasoned with seafood-based salinity in a way that will bore a clear hole through your foggy head. I actually keep a bag of kongnamul in my fridge on standby because I make this soup all the time, regardless of hangover status.

KONGNAMUL GUK (BEAN SPROUT SOUP)
콩나물국

Serves 4

1 (6-inch) piece dasima (dried kelp)/kombu

8 large dried anchovies (for broth)

2 dried shiitake mushrooms

1 (12-ounce) bag kongnamul (soybean sprouts)

1 tablespoon minced garlic

1 tablespoon salted shrimp

1 tablespoon soup soy sauce (see page 221)

½ teaspoon gochugaru

2 scallions, chopped

2 cups fresh or frozen thinly sliced squid or frozen seafood mix (no need to thaw)

½ to 1 Cheongyang chili pepper (optional)

Roasted seaweed, for garnish

Cooked rice, for serving

1. In a pot, combine the dasima, anchovies, mushrooms, and 2 quarts water. Bring to a boil over high heat. Immediately reduce to a simmer and remove the kelp, which you can save for another use (see "Cold Brew" Leftover Dasima Broth, page 121). Simmer the anchovies and mushrooms (there should be some bubbles as they cook), partially covered, for at least 15 and up to 30 minutes.

2. Remove the anchovies and mushrooms, the latter of which you can also save for another use. Increase the heat to medium and add the soybean sprouts and minced garlic. Cook for 15 minutes, covered or uncovered. (Choose one and stick with it! See Cooking with Kongnamul on page 212 for an explanation.)

3. Stir in the salted shrimp, soup soy sauce, gochugaru, scallions, squid or seafood mix, and Cheongyang chili pepper (if using). Bring back up to a simmer and cook for a few minutes longer so the flavors meld. Taste and adjust seasonings as desired.

4. Crush lots of roasted seaweed atop the soup and serve with rice and copious amounts of hydration. Take a nap when you're done; you deserve it.

COOKING WITH KONGNAMUL

Washing soybean sprouts can be a bit of a pain; it's one of the things my mom most likes to nitpick about when I'm cooking with her. Here are my tips, based loosely on hers, but note that, to be honest, as long as the bean sprouts don't smell or look too bad, you don't have to be that perfect with them.

• **WASHING:** Put the sprouts in a sieve placed inside a bowl and rinse with water three times, picking out any loose bean skins and tearing off any browned tips.

• **COOKING:** Soybean sprouts can get kinda fishy if cooked incorrectly, and the general old wives' tale is that you need to cook them with the lid on, or with the lid off, and you can't switch partway through. I personally like to cook them with the lid off so I can see how they're doing without the stress of "Oh no, is it bad to open the lid now?"

• **STORING:** I love having soybean sprouts on hand because they are exactly what I crave when I'm hungover, but their shelf life is unfortunately short. If packaged, keep them in the bag they came in until ready to use (don't use them if too much brown water has accumulated at the bottom). If you open the package but don't use the whole bag, treat them like you would tofu: Store in a container with clean water, changing daily, and use them within the next few days.

You might wonder why there's yet another recipe for instant ramyun in this book (see page 101 for the other). The package comes with its own directions, after all. But to be honest, it's what I crave when I'm hungover. (Also when I'm drinking. Also after I've been traveling. And when I'm traveling. Maybe I just crave ramyun all the time.) But the key to this instant ramyun is that it's chock-full of ingredients specifically targeted at alleviating your hangover, like bean sprouts, kimchi, and egg. It'll make you sweat out the night before, fill up your belly, and make you nice and sleepy for another nap.

This recipe does not contain exact measurements, because let's be real: When you're feeding your hangover with ramyun, it's because you can't be bothered with anything exact. Use what you have in your fridge, sub where needed, or, when all else fails, just dump it all in and call it a day (back to bed!).

1. In a small pot heat a bit of oil over medium heat. Add the scallions and stir-fry a minute or two, until fragrant.

2. Add the kimchi, gochugaru, and ramyun seasoning packet and stir to combine. Add water for the ramyun: a little more than the package calls for (2½ cups, or 600 ml if you want to be more exact). Bring to a boil over high heat, then add the ramyun flakes and bean sprouts. Add the ramyun noodles and cook for 2 minutes.

3. Add the chili pepper and crack in the egg. Cook for another 2 minutes or until the egg is to your desired consistency. I like to agitate the white a little to cook faster in the hot broth and let the yolk poach. Eat and be restored.

HANGOVER RAMYUN
해장라면

Serves 1

Some oil

A scallion or two, chopped (it's a good time to use those wilted ones at the bottom of your crisper)

The dregs of your kimchi jar (¼ cup, about 2 chopstickfuls)

½ spoonful of gochugaru

Your favorite ramyun (my choice is the classic Shin Ramyun)

A large fistful of soybean sprouts or mung bean sprouts

A bit of Cheongyang chili pepper, jalapeño, or serrano, thinly sliced

An egg

GRILLED KIMCHEEZE
김치 그릴드 치즈

Makes 4 sandwiches

1 tablespoon unsalted butter

1 small yellow onion, thinly sliced

Pinch of kosher salt

⅓ cup chopped kimchi

1 plum tomato, diced

SANDWICHES

8 slices (½ inch thick) sourdough bread

Mayonnaise, Kewpie or regular

2 tablespoons whole-grain mustard

1⅓ heaping cups grated cheddar cheese

6 slices Kraft American cheese

Some hangovers need to be cured with hot, soothing soup; other hangovers call for greasy, carb-y deliciousness. This grilled cheese sandwich is for when you desperately need the latter to soak up all the lingering booze sloshing around in your stomach from the night before.

Make a batch of the onion-kimchi-tomato mixture in advance when you have some (sober) free time. Cooking it can take a while, but it's well worth the extra effort: The onions and kimchi get nice and rich while the tomatoes stay bright and acidic. You can use any leftovers whenever you feel like it: Swirl into some scrambled eggs, or mix with some warm rice.

1. In a cast-iron skillet, melt the butter over medium-low heat. Add the sliced onion, season with the salt, and cook, stirring occasionally, until the onions are melty and lightly caramelized, about 20 minutes.

2. Add the chopped kimchi and cook, stirring occasionally, until the moisture from the kimchi evaporates and it becomes lightly charred on the edges, about 10 minutes.

3. Stir in the diced tomato and cook for 2 to 3 minutes to evaporate the moisture (the tomatoes won't be totally cooked, but that's fine). Remove the mixture from the heat and set aside.

4. Prep the sandwiches Place sourdough slices on a cutting board and spread each slice with some mayo. Flip half the slices over and spread these with the mustard. Lay 1½ slices American cheese on each mustard side of the 4 sourdough slices. Divide the onion-kimchi-tomato mixture and the cheddar evenly over the cheese. Place the other 4 slices on top mayo-side up and press down gently.

5. Reheat the cast-iron skillet over medium heat (or preheat a panini press). Nestle one or two sandwiches in depending on the size of your pan. Grill until the bottom of the sandwich is golden brown, 4 to 5 minutes. Flip and cook another 1 to 2 minutes until the cheese has melted (you can put a lid on the pan to speed up the process, but don't let the bread burn). Repeat with the remaining sandwiches.

6. Slice the sandwiches in half in whichever direction tickles your fancy and serve immediately.

Pantry

Koreans have long been adapting their food to the terrain at hand as well as to their lifestyle, so I'd like to think that nearly any ingredient in these recipes can be substituted with something else you have on hand or is available at your local grocery store. It's totally fine to use the soy sauce that's in your cupboard or skip the homemade broth in favor of instant. But if you want to stock up your pantry, below is some guidance on how and what to shop for.

CIDER 사이다

All the cider referred to in this book is Korean cider, specifically 사이다, which is a lemon-lime soda, not apple cider. I most often use Chilsung Cider, which has a bright, balanced, fizzy sweetness, but if you can't find it you can also use Sprite or 7Up.

COOKING SYRUPS 시럽

Personally, I find the Korean sweeteners fairly interchangeable—they will all get the job done (adding sweetness and helping to thicken sauces or marinades) with small differences depending on the dish. For this book, you can generally buy one of the following and use throughout. I most prefer rice oligodang 올리고당 (malted rice syrup) for its balanced depth and sweetness, but you can go more malty with jocheong 조청, or less malty with ssalyeot 쌀엿 (rice syrup) or mulyeot 물엿 (corn syrup). Or, if you're in a pinch, sub with honey or sugar, though you'll have to adjust the amount and the water level for the latter. I won't judge you, even if your mom might.

DASIMA 다시마 (KELP)

Also known as kombu, this is sold in large dried pieces at Asian grocery stores. Buy in bulk and cut into 6-inch-long pieces; store indefinitely in a ziplock bag. This will be the perfect amount for each recipe. I also like to reuse spent kelp for a second "cold brew" broth (see page 121) or as a stand-alone banchan.

DDUK 떡 (RICE CAKES)

Korean rice cakes come in disk form (used for soups) or log form (used for ddukbokki). Freshly made is best (if you can find it at your local Asian grocery store or dduk shop), but you should also be able to reliably find them in the refrigerated section. Look for rice cakes made with minimal ingredients (just rice, salt, and alcohol as a preservative). They will keep for several weeks in your fridge—you can freeze them for long-term storage if you wish, but sometimes the rice cakes will start to split if frozen for too long or defrosted too many times. Soak the rice cakes in water for ten to thirty minutes prior to use, to get rid of the alcohol preservative and to break up any pieces that might be stuck together.

DOENJANG 된장

This fermented soybean paste is often used as the basis for stews and braises. Look for a doenjang without additives; I particularly like Baekil 백일된장 (meaning "100 days," how long this doenjang was fermented and aged) or Tojang 토장, both from Sempio.

GANJANG 간장 (SOY SAUCE)

All recipes that call for soy sauce have been developed with Sempio Naturally Brewed Soy Sauce 501 샘표 양조간장 501, unless it calls for soup soy sauce (see next page). There are so many types of soy sauce at the Asian grocery store, it's hard to know where to start deciphering the differences. After a deep dive into the history and the diversity of these Korean soy sauces, here are my preferences when cooking these recipes.

Yangjo Ganjang 양조간장 (Brewed Soy Sauce)

This is the Korean term for the more common soy sauce, which is usually brewed with soybeans and wheat. It differs from jin soy sauce 진간장, which is mixed with a more chemically produced soy sauce. I usually reach for Sempio Naturally Brewed Soy Sauce 501 샘표 양조간장 501, which has a balanced flavor that's nicely salty with a touch of sweetness. You can use whatever soy sauce you have on hand for these recipes (Lee Kum Kee Premium Soy Sauce and Kikkoman Less Sodium Soy Sauce are good substitutes), though the salinity may differ across products and you should adjust your seasoning to taste.

Guk Ganjang 국간장 (Soup Soy Sauce)

This is a specifically Korean-style soy sauce that's created as a by-product of the doenjang-making process, so it has a nice strong salty and fermented quality that's hard to replicate. I use Sempio Chosun Naturally Brewed Soy Sauce for Soup 샘표 조선간장, but if you can't find this at your local Asian grocery store, you can substitute a mix of regular soy sauce plus additional salt or fish sauce.

GOCHUJANG 고추장

Gochujang, a fermented chili paste, comes in different spice levels, ranging from mild to very hot, or on a scale from 1 to 5. I would start with a 1 or 2 (mild or medium), even if you are very pro-spice, because with the quantity used in cooking, the heat will blow out your palate.

MYEOLCHI 멸치 (DRIED ANCHOVIES)

Dried anchovies come in a few different sizes: the tiny ones and the medium ones are stir-fried into myeolchi-bokkeum (see page 65) while the larger ones are used for broth. For the latter, look for ones that have shiny, silvery skin and bright round eyes. Before cooking, use your thumb to break open just under the head and remove the black guts (known as *poop* 똥 in Korean).

PRODUCE

Korean Peppers 고추

Korean peppers can be a little hard to decipher at first glance. The slightly longer green peppers served at Korean BBQ restaurants alongside cucumbers or carrots with a dollop of ssamjang are oi-gochu 오이고추, or cucumber-style peppers, that are not supposed to be very spicy. Tread carefully, though, and take a little nibble to start, as they can be way hotter than you think.

Cheongyang Chili Pepper 청양고추

On the flip side, the smaller (but also green) peppers are usually Cheongyang chili peppers, which *are* spicy. There's a range of spice on these guys as well, so even if you are a spice-lover, start with half a pepper and add the rest after tasting. You can also remove the seeds to temper the heat level. In this book, I call for a range when adding Cheongyang chili peppers, for exactly this reason. Look for these peppers at the Korean grocery store or sub with serranos or jalapeños.

Mu 무 (Korean Radish)

This stout white radish is commonly used in making kimchi and lends a deep flavor to soups. If you can't find it, daikon is a fine substitute, but other radishes are not.

Perilla Leaves 깻잎

Also sold as sesame leaves, these are much larger and grassier than their Japanese counterpart, shiso. You can find these at Asian grocery stores, sold in small bundles; if you have leftovers from a recipe, use as a wrap for Korean BBQ or julienne to use in a salad.

Scallions 파 (Green Onions)

All recipes in this book call for American-style scallions; look for firm, medium ones. But in South Korea there are a few different types of green onions. Daepa 대파 is a larger green onion that's similar to but smaller than a leek (it's also known as a calçot in Spain). If you find these in grocery stores, you can wash, chop, and freeze them to cook into soups and stews. Silpa 실파 is most similar to the American scallion, but is harvested when it's young, thin, and tender, while jjokpa 쪽파 has a more pronounced bulb end; both of these can be used to cook with and for garnishing. Try subbing in one of these Korean-style green onions if you can find them, for a slightly different texture and flavor.

RICE 쌀

Korean rice is a medium-grain white rice. You can usually buy it in 5-pound, 15-pound, or 40-pound bags at an Asian grocery store. I like Kyong Gi Me 경기미 rice, as well as Kokuho Rice (a similar, California-grown variety). Any Japanese or Calrose rice will do; in a pinch you can also use "sushi rice," though it tends to be quite expensive for a small quantity. Cook rice fresh in a rice cooker or on the stovetop, or stock up on microwavable single-portion cooked rice like Hetbahn 햇반.

SALT 소금

All recipes have been developed with Diamond Crystal kosher salt. If using Morton kosher salt, sea salt, or table salt, halve the amount of salt, and add more to taste.

YUJA / YUZU JUICE 유자즙

This can be found at Japanese specialty stores or online. Yuzuco is a great newer brand that sells 100% yuzu juice.

Tools

While this isn't an exhaustive list of tools you might need to make these recipes, here are a few specialty items that might be worth your time to go out of your way to source.

JIGGER

A jigger is a bar tool used to measure liquid for cocktails. I like to use a bell jigger, a curved, double-sided vessel, the larger side of which measures 2 ounces and the smaller side of which measures 1 ounce.

KITCHEN SCISSORS

I could not imagine cooking in a kitchen without these, which is why I have multiple heavy-duty kitchen scissors stashed at home, the bar, and even my mother-in-law's house. I use them to open bags, cut meat when grilling, break down whole chickens, or snip scallions for garnishes.

KOREAN BBQ TONGS

These stainless steel tongs exist in the space between the larger unwieldly American BBQ-style grill tongs and tiny kitchen tweezers. While they're made for flipping thin slices of meat at Korean barbecue, I use them for everyday cooking and serving.

MUSUBI MOLD

A unitasker, sure, but these plastic molds are invaluable for making musubi at home. You can easily find one online or at Asian grocery stores for pretty cheap.

NONSTICK WOK OR SKILLET

While Koreans may not have a cooking pan that's as iconic in modern kitchens as the carbon steel wok may be for the Chinese, I'm partial to the ever-versatile, incredibly accessible nonstick wok. Its high, rounded, sloped sides make for easy large-format cooking, and cleanup is much less fraught. There's a reason my mom always reached for this in her kitchen, and why it's been a mainstay in mine.

SOJU GLASS

This refers to a 2-ounce shot glass and is slightly larger than your traditional American 1½-ounce shot glass.

SOMAEK GLASS

This is an 8-ounce glass that Koreans use for beer, and also for making Somaek (page 4).

Acknowledgments

My mom doesn't drink, but I clearly do.

If you know anything about Korean moms, you know that their approval (and disapproval) means a lot, even once you've become a grown, legally drinking adult, and so I assumed she'd never approve of my silly drunken antics. One night my mom sent a video she saw online of me pounding a Milkis shot that someone else had posted. It had caught the algorithm train and somehow ended up on her feed. I braced myself for a chastising. Instead, she asked, "Is this you? Why aren't you tagged in this video? Shouldn't you be getting credit?"

So the next morning I posted the video to my account, and "How to Make a Milkis Shot" went on to garner more than 16 million views and counting. It made me realize that people were curious and eager to learn about Korean alcohol, food, and drinking culture. This led to a series of events that led to this book, so I want to express my thanks to:

Madison Trapkin and Erin Alexander, for assigning me to write a piece about Korean drinking traditions for Food52.

Claire Yee, for being the first to believe that this idea could be a book, which until then had just been a "maybe one day" dream of mine.

Danielle Svetcov, for becoming my agent and being straightforward, generous, and kind with your guidance, not only to me but also to everyone else around me.

Christina Chaey, for coming up with the name of this book.

Tom Pold, for your thoughtful and calmly encouraging edits and support, and for instilling trust in me that I could write this book in my own way (which was a scary thing to do).

Heami Lee, I believe that a cookbook lives and dies by its photos, and I'm so honored that you breathed this much life and infused so much of our culture into mine.

Eugene Jho, for going so above and beyond in sharing your space, your legendary food-styling skills, and your genuine enthusiasm. I feel so blessed to have worked with you on this.

Maggie DiMarco, for your pitch-perfect patterns and props, and for incorporating all my random Korean pop culture references.

Jean Lee, for helping make this photoshoot happen.

Alyssa Kondracki, Veronica Martinez, Ava Chambers, James Park, Sam Kang, and Christina for your help and hands on the shoot.

Alex Distell, Ashley Noh, Justine Lee, Trent Pheifer, and James, and Christina, again, for ensuring that these recipes are tested and tasty.

The team at Knopf, including Anna Knighton, Kelly Blair, Meredith Dros, Hilary DiLoreto, and Melissa Yoon, for making this book possible and perfect.

And special thanks to a few important people:

희대야—네가 소주병 돌리는 법을 가르쳐 줘서 내가 책 썼다. 그날 밤에 톡을 보내줘서 고맙다.

희정아—언니가 소주 연구를 할 수 있게 한국에서 책들을 들고 와줘서 넘 고마워~

To Peter and Patty, for your immediate acceptance and love, and for your unwavering support and patience.

To Jamesy, it was in my kitchen that you embarked on your cookbook journey. I'm so grateful that you've been a part of mine, and thank you for always sharing your support, encouragement, feedback, and space. 우리 계속 같이 파이팅 하자!

To Carolyn, thank you so much for your amazing illustrations—I'm truly in awe of how perfect they turned out. And, of course, thanks for always being my sounding board, gut check, and forever sister champion. I'm so happy that you are a part of this book and that we can share a title page together.

To Nick, thank you for doing dozens of pop-ups with me, for your discerning palate, for your enlightening ideas, and especially for sharing your cocktail recipes in this book without asking for a single ounce of credit (here it is, though). Thanks for being my number one fan, and for being my partner in life, in business, and in love. Thank you always for taking such good care of me and our family (and for being the one who actually cooks).

엄마 아빠에게—많은 사랑과 도움을 주셔서 감사하고 맛있는 것들 많이 매겨주시고 좋은 경험도 많이 주셔서 고마워요.

그리고 우리 제하야—있어줘서 고마워.

And for the record, the last time my mom showed this much interest in something I was doing, it was my test scores in high school. So I think she definitely approves of this book. I hope you do, too.

Index

(Page references in *italics* refer to illustrations.)

Truffle Jjapagetti with
    Beef Brisket, *100*,
    101
twist cap game, *178*, *180*, 181

U

Ugeoji, 208
Union Pool, Williamsburg,
    Brooklyn, 71

V

Van Brunt Stillhouse, xli
vermouth:
    dry, in Naengmyeon, *136*,
        137
    sweet, in Jujube Ginseng
        Negroni, *116*, 117

W

Watermelon Soju
    Hwachae, *150*,
    151
West 32 Soju, xli
where Koreans drink,
    lviii–lix
whiskey:
    Misugaru (cocktail),
        130, *131*
    see also rye
        whiskey
A Whiskey Sour
    (Fall Story), 114, *115*
White Ddukbokki,
    90, *91*
wok, nonstick, 225
wonju, xxxiii
Wonny's Bok-So-Ju,
    36, 37
Won Soju, xli

wraps:
    Bossam (Boiled Pork
        Wraps) with Garlic
        Chive Sauce, *92*, 93
    ssam, building, 159

Y

yakju ("medicinal wine"), lxiii,
    lxv
Yakult, 138, *139*
Yangchon Brewery, lxiii
yangjo ganjang (brewed soy
    sauce), 221
Yipseju, xxxvii
Yobo Soju, xli
yogurt drink (Yakult), 138
yuja (yuzu) juice, 223
    Perilla Yuja Soju & Tonic,
        106, *107*
    Yuja Michelada, 110, *111*

A NOTE ON THE TYPE

This book was set in Archer, designed by Jonathan Hoefler and published by Hoefler&Co. in 2001. While rooted in the geometric slab serif style that emerged in the early nineteenth century, Archer's many liberties with the style include its uncharacteristic application of ball terminals to capital letters such as C and G, a detail traditionally found only in the lowercase alphabet. For a slab serif, it has a very warm and inviting feeling to it.

Composed by North Market Street Graphics, Lancaster, Pennsylvania

Designed by Anna B. Knighton

"*Soju Party* is an invitation to accompany Irene Yoo and her friends for an epic night out, which spills into the haze of a hangover the morning after. The organization of the book is joyously deployed to situate contemporary Korean food and drink recipes, historical anecdotes, pop cultural insights, and idiosyncratic hospitality traditions in their ideal venue. At a time when alcohol is being villainized (again), I'm thrilled to see a no-holds-barred guide that shows how soju and other Korean beverages function as a lubricant that reveals and holds together kaleidoscopically complicated social structures gleefully under the artful eye of an exuberant host."

JIM MEEHAN,

author of *The Bartender's Pantry*,
Meehan's Bartender Manual, and *The PDT Cocktail Book*

"More than just a cookbook, this is the ultimate Korean drinking experience, with Queen of the Soju Tornado Irene Yoo as your guide. In Korean culture, drinking is about bonding . . . and eating, and by the time you finish this book, you'll want to throw your own soju party!"

JAMES PARK,

author of *Chili Crisp*, @jamesyworld